Conspiracies in Religion, Politics and Science

Conspiracies in Religion, Politics and Science

Contents

1 Prologue 11
    1.1 Why Russians Never Landed a Man on Moon? 11
2 Chapter 02 15
    2.1 Background to Free Inquiry 15
        2.1.1   Sir John Kotalawala 18
        2.1.2   Narada Maha Thera 21
        2.1.3   Power by Bertrand Russell. 24
        2.1.4   Nyanatiloka Mahathera 27
            2.1.4.1   World War I 27
            2.1.4.2   World War II 27
        2.1.5   Nyanaponika Thera 28
3 Chapter 03 33
    3.1 Blind Logics and Logistics of Modern Society 33
4 Chapter 04 39
    4.1 Why Philosophy? 39
    4.2 Hippocrates 39
    4.3 Plato 40
5 Chapter 05 43
    5.1 Aristotle 43
6 Chapter 06 45
    6.1 Religion 45
        6.1.1   Why the decadence? 45
        6.1.2   What contribution the religion has made? 45
7 Chapter 07 47
    7.1 Meaning of Freedom 47
    7.2 Freedom 55
    7.3 Freedom Revisited 56
8 Chapter 08 59
    8.1 Political Wisdom 59
    8.2 Reflection of our Polity 62
9 Chapter 09 67
    9.1 Basic Tenet of Human Civilization 67
        9.1.1   Galactic Civilizations 68
        9.1.2   Visit by Advanced Civilizations 68
        9.1.3   By Nigel Calder and John Newell 69
    9.2 Paul Hellyer, Canadian Minister of Defense. 71

- 9.2.1 Extraterrestrial Issues.......... 73
- 9.3 The Day After Roswell............................... 75
  - 9.3.1 Philip J. Corso..................... 76
  - 9.3.2 General George C. Marshall.78
  - 9.3.3 Vice Admiral Roscoe Hillenkoetter........... 78
    - 9.3.3.1 NICAP...................... 79
  - 9.3.4 Victor Marchetti................. 80
  - 9.3.5 J. Edgar Hoover, head of the FBI....... 81
  - 9.3.6 Harry Truman...................... 81
  - 9.3.7 President John F. Kennedy in 1963.......... 82
- 10 Chapter 10 83
  - 10.1 Insight.. 83
  - 10.2 Mental Culture... 84
- 11 Chapter 11.. 87
  - 11.1 Imagination.. 87
- 12 Chapter 12.. 89
  - 12.1 Conversion... 89
- 13 Chapter 13.. 91
  - 13.1 Basic tenet of Science............................... 91
- 14 Chapter 14.. 95
  - 14.1 How We Killed Our Science Education...... 95
- 15 Chapter 15.. 97
  - 15.1 Why Science took a back seat in Human History........ 97
  - 15.2 Right Constitution..................................... 98
- 16 Chapter 16.. 99
  - 16.1 Conspiracies in Science............................. 99
    - 16.1.1 Microsoft's Strategy............ 102
- 17 Chapter 17.. 105
  - 17.1 Drug Companies... 105
    - 17.1.1 Chocolates and Drugs.......... 105
  - 17.2 Late Professor Senaka Bibile..................... 107
    - 17.2.1 Cholesterol and its treatment. 110
- 18 Chapter 18.. 113
  - 18.1 Zoonosis... 113
  - 18.2 Colonial Expansion 115
    - 18.2.1 Exploitation and atrocities... 117
    - 18.2.2 Mutilated people from the Congo Free State... 118
    - 18.2.3 Medicine, Science and Germans in Mid West Africa... 118
  - 18.3 AIDS Epidemic.. 119

- 18.3.1 Bushmeat Hypothesis 123
- 18.3.2 Origin and the Epidemic 127
- 18.3.3 Colonialism in Africa 128
- 18.3.4 Unsterile injections 129
- 18.3.5 The injection campaigns against sleeping sickness 130

19 Chapter 19 133

19.1 UFOs 133
- 19.1.1 Project Sign 135
- 19.1.2 Project Grudge 136
- 19.1.3 Condon Committee 137
- 19.1.4 USAF Regulation 200-2 138
- 19.1.5 Project Blue Book 138
- 19.1.6 CUFOS 140
- 19.1.7 MUFON 140

19.2 The Soviet UFO Sightings 142
19.3 UFOs, Area 51 and Conspiracy 143
19.4 Center for the Study of Extraterrestrial Intelligence 145
19.5 Disclosure 147
- 19.5.1 Brian Todd O'Leary 148
- 19.5.2 Steven Macon Greer 152
- 19.5.3 John Callahan 153
- 19.5.4 The Anchorage Incident 154
- 19.5.5 FAA Reporting 155
- 19.5.6 Other Sightings 156
- 19.5.7 Stephen G. Bassett 156
- 19.5.8 John Lear 157
- 19.5.9 Richard C. Hoagland 158
- 19.5.10 Milton William "Bill" Cooper .. 159
- 19.5.11 UFOs, Aliens and the Illuminati 159
- 19.5.12 Stanton T. Friedman 163

19.6 Friedman's positions regarding UFO phenomena 164
- 19.6.1 Barney and Betty Hill 165
- 19.6.2 UFO encounter 165
- 19.6.3 The Star Map 170

19.7 Development in 1980s 172
- 19.7.1 MJ-12 172
- 19.7.2 Linda Moulton Howe 172
- 19.7.3 Milton William Cooper 172
- 19.7.4 Bob Lazar 172

19.8 USA / UFO Scenario 174

- 20 Chapter 20 179
  - 20.1 Big Bang Conspiracy 179
- 21 Chapter 21 183
  - 21.1 Alternative Theories to "Big Bang" 183
    - 21.1.1 The Incredible Bulk 184
    - 21.1.2 Time's Arrow 189
    - 21.1.3 The Nows Have It 196
- 22 Chapter 22 201
  - 22.1 Global Warming 201
  - 22.2 Population Expansion and Global Warming. 201
    - 22.2.1 Forest Harvesting 203
    - 22.2.2 My Exotic Tropical Fish 204
- 23 Chapter 23 209
  - 23.1 Mad Cow Disease 209
- 24 Chapter 24 213
  - 24.1 Coconut Theory 213
    - 24.1.1 Coconut Experience 214
    - 24.1.2 PCBs 214
    - 24.1.3 PCBPs 215
  - 24.2 A Case for Coconut oil 217
  - 24.3 Coconut Saga 220
    - 24.3.1 How to prepare Virgin Oil at home 227
    - 24.3.2 Traditional Way 227
    - 24.3.3 My Method 228
- 25 Chapter 25 231
  - 25.1 Chip Industry 231
  - 25.2 OEM 231
    - 25.2.1 The Raspberry Pi 233
  - 25.3 Jack Kilby's original integrated circuit 237
  - 25.4 Intel Corporation 239
  - 25.5 Advanced Micro Devices (AMD) 240
  - 25.6 IBM 241
    - 25.6.1 Fabless 242
  - 25.7 DOS 243
  - 25.8 MS-DOS 244
  - 25.9 Microsoft 245
  - 25.10 Linux 247
  - 25.11 Richard Matthew Stallman 249
  - 25.12 Linus Benedict Torvalds 251
    - 25.12.1 When Linus Torvalds Almost Quit 254
  - 25.13 Jon Hall 256

- 25.14 Ian Murdock.................................... 258
- 25.15 Klaus Knopper............................... 262
- 25.16 Peter Parfitt.................................... 263
- 25.17 Unix.. 263

26 Chapter 26.. 265
- 26.1 Solar Cells... 265
- 26.2 Alexandre-Edmond Becquerel................... 267
- 26.3 Space applications................................ 268

27 Chapter 27.. 269
- 27.1 Lithium Dry Cell battery........................ 269
- 27.2 History Electric Cell............................... 269

28 Chapter 28.. 277
- 28.1 Petroleum Oil Conspiracy....................... 277
- 28.2 History of Fuel Oil................................. 277

29 Chapter 29.. 283
- 29.1 Linux.. 283
- 29.2 History of Linux.................................... 285

30 Chapter 30.. 291
- 30.1 Myths and Conspiracies in Human History. 291
- 30.2 Conspiracies in Science.......................... 293

31 Chapter 31.. 295
- 31.1 Emerging Science-01.............................. 295
  - 31.1.1 Elliptical galaxies................ 295
  - 31.1.2 Spiral galaxies..................... 295
  - 31.1.3 Starburst or Irregular Galaxies... 296
  - 31.1.4 Galaxy Merger..................... 296
  - 31.1.5 Dark matter within galaxies. 296
- 31.2 Emerging Science-02.............................. 300
  - 31.2.1 This is my alternative theory to counteract the "Big Bang", the Holy One?... 300
- 31.3 Emerging Science-03.............................. 304
- 31.4 Emerging Science-04.............................. 308
  - 31.4.1 Zipping and Unzipping Model gives Flexibility and Modularity to Dark Matter.. 308

32 Chapter 32.. 309
- 32.1 Astrology, Astrophysics and switching of Time Dimension... 309

33 Chapter 33.. 315
- 33.1 Laser and Holography and their Uses........ 315
  - 33.1.1 Interference pattern 320
  - 33.1.2 Transmission hologram........ 322

- 33.1.3 Transmission hologram recording........ 323
- 34 Chapter 34... 325
    - 34.1 Conspiracies in Politics-01.......................... 325
        - 34.1.1 Why we need a Referendum to get rid of the Need of a Referendum to make Constitutional Effects............................ 325
    - 34.2 Conspiracies in Politics-02.......................... 328
        - 34.2.1 Franchise a mathematical marvel or an aberration of Sri-Lankan Polity..... 328
    - 34.3 Conspiracies in Politics-03.......................... 332
        - 34.3.1 Verse 324, 320 and 321 of Dhammapada... 332
        - 34.3.2 The Elephant Corridor......... 333
- 35 Epilogue... 337
    - 35.1 Henry Steel Colonel Olcott.......................... 337
        - 35.1.1 Theosophical Society 337
    - 35.2 Helena Blavatsky... 338
    - 35.3 Annie Besant.. 340
    - 35.4 Buddhism in America.................................. 342
    - 35.5 Rhys Davies.. 343
        - 35.5.1 British Conspiracy............... 344
        - 35.5.2 Looking for absolute truth... 345
    - 35.6 Prof ADP Kalansuriya................................. 346
    - 35.7 Authors Note.. 349

Prologue

Why Russians Never Landed a Man on Moon?

Following hypotheses of mine are outrageous but nevertheless needs probing. I am not going to include them in this book.

Human conspiracy is in the genes, and outlining it is an uphill task.

I have looked at, only the scrapings.

It is the tip of an iceberg.

What is hidden below is awful and appalling.

I do not intend to reveal all of them and waste rest of my life on a failed endeavor.

I have better things to do in my life.

Imagination, insight and conviction are vital and necessary ingredients to outsmart conspiracy from its very base. Conversion without contradictions, leads one to the trappings of all human conspiracies.

Coming to my pondering of Russian actions.

1. Russia did not have the technology.
This may not be true.
They were well ahead of America at a particular period.

2. Russia did not have the money.
Probable but cannot be substantiated.

3. Russia was prudent not to waste their scarce resources.

This is true in a communist but relatively poor country.

4. Russia did not want to sacrifice the lives of their trained astronauts for an ego exercise.

This is my bone of contention.

They were more useful in defense strategies.

5. None of the above.

Most likely scenario.

Now let me take the American scenario.

1. American public are generally stupid (mind you not all) and are mostly afflicted by an incurable ego problem.

This is true.

2. They want to be the first in scientific advances and pay or hijack (Germans after World War II) personnel to man their projects.

Another way how ego manifests.

3. They had enough resources but wasted them on putative communist war.

True.

4. They had a powerful secret service and they did not want politicians to take credit.

Yes.

5. JFK was assassinated to channel money (not for space program) to CIA projects.

Yes.

6. America never send any astronauts to the moon (to save money for covert operations) but choreographed either by holography or actual photography on earth.

This is my belief even though there is no proof available.

I am human and have the freedom for skepticism.

7. America did not have a correct sense or directive with so many lobbies to contend with.

Really?

8. They had to satisfy the Vatican.

Most likely overtly or covertly.

9. None of the above.

Most likely scenario.

Now my revelation hypothesis.

1. Russia and USA did not collaborate with their defense policy.

No.

In fact, they cooperate.

2. They do not want conflicts to throttle their own programs.

Yes.

3. They are in contact with aliens and are using their technology (without the knowledge of the public).

Yes.

4. Russia, America and Aliens already have an advanced interplanetary mission.

Most likely.

5. America (CIA) is using drug money for the clandestine black projects, since the Legislative Assemblies are very slow to authorize money or veto them outright.

Chapter 02

Background to Free Inquiry

The idea of this book was in my mind for sometime but deliberately delayed it, so that I could devote much time and effort on the final outcome. The idea was to have an overview of the human thinking over the last 3000 years based on written or oral material of the East and West. How science shaped and developed my way of thinking is altogether a different kettle of fish.

I must say, I had the luxury of free thinking from my childhood and nobody tried to modify it.

In other words, I did not have a mentor.

Except for the sports master, in my simple village school who instilled in my mind, the golden rule in sports "It is participation, not winning that matters" under the Olympics banner and the late Professor Senaka Bibile in my undergraduate years, there were only a very few personalities that I admired.

An exception to that rule was Sir john Kotalawala whom I used to see as a tiny tot. He was famous for his uttering "It is better, go fox hunting in England than do politics in Ceylon". Shooting foxes was a sport in UK then, I joined the cavalcade, vigorously opposing it, as a sport, in early eighties, on the lines of animals rights, while working there.

We won our campaign and it was abolished on the grounds of preventing cruelty to animals. Unfortunately in this country we still cause untold cruelty to our "Gentle Giant", a political symbol of a major party in the name of traditional Buddhist practice of parading them in procession.

Some monks of our tradition are the exponent of this "cruelty sport" while illegally hunting the baby elephants, in the wild joining hand with Ganja (hashish) growers.

Having said that, I did not understand the deeper meaning of Buddhism (till later years in my life), but its principle practices have had an influence in my thinking.

It teaches us of simplicity and despise grandiosity.

Avihinsa (nonviolence), the guiding principle, which is an antithesis to our polity including priests.

For a starter, I did not believe in creation of the world by a single god. That meant I did not believe in existence of any form of god dead or live.

Advent of the Russian exploration in space when I was a tiny tot, not only made deeper meaning and admiration of its people without a religion.

I must say, I did not want the communist pill down my throat (like a bitter pill).

In adult life, I revised my view of Russia significantly (not antagonistically). Just imagine, how nasty it would have been, if we were bulldozed by America (and the West) only.

For my luck, I was born with scientific outlook and learning science was a pleasure (language and religion were not) and a pastime.

Add to that there were plenty of reading materials, including local papers and scientific journals at the British

Council Library. English language or literature (I hated Shakespeare but made a point to visit his birth place, when I was working in UK) was not my forte but that did not deter me from grasping scientific terminology.

In other words, the outside world contributed a lot to my formative years.

The fact, that I was born to an independent, free nation was a feather on my cap.

There was no indoctrination and freedom to think and act were bonuses.

How, I went up the ladder was a shear chance occurrence with no prior programming.

The rationalist Abram Kovoor was making his presence felt and I had the audacity to pose probing question for him to answer. I never followed his line of thinking but admired him all the same.

Mind you, human way is to justify and rationalize any bizarre practice or conviction from prehistoric times. They even go to the extent of sacrificing life of even children to the devil (Bahiravaya) or god.

Abraham Thomas Kovoor (April 10th,1898- September18th, 1978) was an Indian professor and rationalist who gained prominence after retirement for his campaign to expose as frauds various Indian and Sri Lankan "god-men" and so-called paranormal phenomena. His criticism of spiritual frauds and organized religions was enthusiastically received by many, initiating a new dynamism in the Rationalist Movement, especially in Sri Lanka. Professor Carlo Fonseka was active somewhat later in time to Kovoor but he did not make an impact on me. He exposed the physiological facts related to fire walking and I did walk on fire (soaked my feet well with water and walked briskly) laid on the stage of the Physiology Lecture Theater.

But I loved his orations and style.

It was a revelation at that time.

A cursory look at the content of this book may look like there is lot of repetition, which I wish I could have avoided, but could not, for the simple reason, the grand conspiracy design was integrated with religion, science and politics.

Sir John Kotalawala

I must confess this is the only Sri-Lankan politician, I am going to mention since he gave up politics in its infancy in Ceylon, and left an indelible mark in my formative brain.

Sir John Kotalawala who was ranked as General in later years was our 3rd Prime Minister of Ceylon from 1953 to 1956. He entered mainstream politics by being elected to the Legislative Council as the member for Kurunegala, thereafter to the State Council in nineteen thirties. In 1948, when Ceylon became a dominion, Kotelawala, was appointed to the Senate. He was elected to the parliament in 1952 and was the Leader of the House and was chosen as the Prime Minister when Dudley Senanayake resigned in 1953.

Sir John was an outspoken man who loved sports and did mix his words in political circles and thereby made a few political enemies in the opposition. He was picked up as a target for slandering and his mavericks made it easy.

Prime Minister Sir John Kotelawala laid the foundation for the "Non Aligned" foreign policy. He took the initiative to hold the South and South East Asian Colombo Powers Conference which became the precursor for the Bandung Conference where the principles of Non Alignment were formulated.

An important happening at the Bandung Conference should be recalled. When the agenda item Colonialism and Independence came to be discussed Sir John raised an issue which caused a stir at the conference.

He actually dropped a bombshell equated Soviet domination over Eastern Europe with Western colonialism in Asia and Africa.

Sir John was always against international communism, Chou-En Lai and Jawaharlal Nehru had got upset and sought to take Sir John to task for his statement but Sir John was said to have, in his inimitable way, maintained his stand.

He was blunt and forthright.

The Indonesia was worried that the conference would collapse but that did not happen and the conference was an absolute success.

Back here in this country the local Marxists had roared and called Sir John "Bandung Boouruwa" but it was the Marxists who turned out to be the absolute Booruwas for when the Iron Curtain fell and the Soviet Union collapsed, the countries of Eastern Europe rose as one to claim how they had been cruelly ruled by Moscow.

In 1996, when Indonesia celebrated an important anniversary of Bandung and our late Foreign Minister Lakshman Kadirgamar was in attendance, he had the onerous task of listening to the heads of Eastern European governments, who were recalling that it was Sir John who had spoken up for them when they were under the Soviet jackboot.

I knew him as a kid and used to visit his Walawwa (Bungalow) with my father. Everybody feared him but he was very easy with children and especially me.

When Queen Elizabeth visited Ceylon in 1953 he invited us and there was a grand party running parallel with the official ceremony.

We were treated like royals.

It was strange coincidence that I had the opportunity to attend to his final illness and make the final diagnosis. I was bit afraid of approaching him and went to him with a senior lady doctor. Of course, he was very nice and he had not lost his charm, even in his twilight years.

Narada Maha Thera

If I do not mention a personality whom I met personally and who had made an impact in my thinking, it is a serious omission on my part. Reverend Narada used to come to the General Hospital, Colombo as an outdoor patient, in his final years.

I used to attend to him and at the end of the procedure, he used to ask me;

What can I do for you, doctor?

I used to say; "Teach me Dhamma, Reverend Sir".

He used to bring me "Bodhi Leaves" published at the Buddhist Publication Society and give me little doses of Dhamma.

One of them was the "Buddhism in a Nutshell".

That was my first introduction to Buddhism in an academic sense.

Later years, I used to go to "Buddhist Publication Society" in Kandy and read all the little publications from Reverend Nyanatiloka Mahathera and Nyanaponika Thera. I never met them in real life.

Buddhism never made an impact on me and I was engrossed in science. I went in search of any reference to "Atomic Theory" in Buddhism. There is a term, "Vibhajja" in Sutta Pitakaya. I had the wrong connotation of the meaning of Vibhajja as "an analytical division".

In Buddhist terminology Vibhajja-Vada; "analytical or discriminating doctrine" is an early name for the original "Buddha

Doctrine". The term Vibhajja-Vadi is not a separate school of thought, but was the characteristic of the Buddha's Teaching;

Now, by blaming what is blamable and praising what is praiseworthy, the Blessed One is a "discriminating teacher" (Vibhajja-Vadi) and is not rigid or one-sided, in his teaching.

When I was searching this term Rev. Piyadhassi Thera happened to be at the BPS. He took the book of Abhidhamma turned the pages and showed me the classified section. Then Rev. Bikku Bodhi, (Editor of the BPS) came around put his hand on my shoulder and said;

"Young man that is not the way to study Dhamma" and disappeared into his room.

What he really meant was don't be academic and do not try to make reference to science in any way. Abhidamma is a colossal text and I did not have the patience to go through it fully. I never searched for any reference to science again but after a bit of research, I found out that the Vibhajja was used in reference to "Mind" and not to "Material".

That was the turning point in my thinking, Buddhism was for the Mind and not for the material science which I was engrossed.

It was strange coincidence that the book, "Power" by Bertrand Russell made a huge impact on me. He was one of the greatest philosophers of the last century.

His writings were colossal.

The keys to human desire, Marx found in wealth and Freud found in sex, (Einstein found in hypothetical particles and forces) but Bertrand Russell instead found it, in power.

Power, he argues, is man's ultimate goal, and is, in its many guises, the single most important element in the development of any society. Writing in the late 1930s when Europe was being torn apart by extremist ideologies and the world was on the brink of war, Russell set out to find a new sense in human thinking.

The result was the Power, a remarkable book that Russell regarded as one of the most important of his long career. Countering the totalitarian desire to dominate, Russell shows how political enlightenment and human understanding can lead to peace - his book is a passionate call for independence of mind and a celebration of the instinctive joy of human life.

Russell claimed that around age 15, he spent considerable time thinking about the validity of Christian religious dogma. He found it very unconvincing. At later stage, he came to the conclusion that there is no free will and that there is no life after death.

Finally, at the age of 18, after reading Mill's "Autobiography", he abandoned his faith and became an atheist.

He wrote a book "Why I am not a Christian?"

This was the final straw. The Church took him to task.

The conspiracy was born.

In early nineteen twenties, he had to be in hiding in either Russia or China. In China he was down with pneumonia. In those days one dies of pneumonia or recover in 14 days. One gets pneumonia on one side or one lobe. Never on both sides. The reason is unknown. Strangely he got pneumonia on the other side (popularly known then, as double pneumonia), no sooner he recovered from the first episode.

He was critically ill.

The Church ran an obituary on him in London and Japan. They were happy that this rouge atheist was gone. To the utter disappointment of the Church he recovered. All due to his lover. Russell's lover Dora Black, a British author, feminist and socialist campaigner, visited Russia, quite accidentally (independently) at about the same time.

She nursed him to good health.

Dora was six months pregnant when the couple returned to England in 1921. Russell arranged a hasty divorce from Alys, the first wife, marrying Dora six days after the divorce was finalized.

The treatment of British and the Church of other faiths is no different, if one looks at, how they treated an atheist.

Power by Bertrand Russell

Russell breaks the forms of influence down into three very general categories;

"the power of force and coercion";

the "power of inducements", such as operant conditioning and group conformity; and

"the power of propaganda and/or habit".

The general effect of an organization, Russell believes, is either to increase the well being of persons, or to aid the survival of the organization itself: "in the main, the effects of organizations, apart from those resulting from governmental self-preservation, are such as to increase individual happiness and well being."

An exhaustive list of the types of organization would be impossible, since the list would be as long as a list of human reasons to organize themselves in groups. However, Russell took only a small sample of organizations to drive his point of view.

The army, the police, economic organizations, educational organizations, organizations of law, political parties and churches are all recognized as societal entities.

The power is in other words is based on Race, Religion, Caste, Creed, Military, Police, Business, Profession (Law, Medicine, Science, Business) and Party.

Even though, I found his English astoundingly difficult in my young age, I somehow grasped the meaning.

There was a subtle insinuation. The conspiracies were built into the Powerful Organizations.

This made me to shun politics all my life.

I think seeds of this book were sown in, by him, indirectly (reading his work), very early in my life.

In addition, I tend to question the validity of science including medicine.

Below, I have given a summary of the contributions of Nyanatiloka Mahathera and Nyanaponika Thera to Buddha Dhamma.

It is pertinent to know how British treated these two monks of western origin during war. It amounts to disdain; if not conspiracy, it was paranoia. It is astounding know that the colossal work was accomplished in solitary confinement or internment.

Bertrand Russel was in prison at least twice.

While in prison, Russell read enormously and wrote the book Introduction to Mathematical Philosophy.

I found prison in many ways quite agreeable. I had no engagements, no difficult decisions to make, no fear of callers, no interruptions to my work. I read enormously; I wrote a book, "Introduction to Mathematical Philosophy"... and began the work for "Analysis of Mind".

I had this book Power which I collected from a bargain sale at the British Council Library, Kandy which had an enormous collection books. Unfortunately, some body had stolen this book when I was abroad. To add, insult to injury, when I returned, the British Council Library Service was abruptly terminated and I had no way of referring to his work.

Currently, it runs as an institute teaching English, as a second language.

Nyanatiloka Mahathera

Nyanatiloka Mahathera (19th February 1878, Wiesbaden, Germany – 28th May 1957, Colombo, Ceylon), born as Anton Gueth, was one of the earliest westerners in modern times to become a Bhikkhu (Buddhist priest), a fully ordained Buddhist monk.

World War I

In 1914, with the outbreak of World War I, Nyanatiloka along with all Germans in British colonies, was interned by the British. First he was allowed to stay at the Island Hermitage, but was then interned in the concentration camp at Diyatalawa, Sri-Lanka. From there he was deported to Australia in 1915, where he mostly stayed at the prison camp at Trial Bay. He was released in 1916 on the condition that he would return to Germany. Instead he traveled by way of Hawaii to China in order to reach the Theravada Buddhist Burmese tribal areas near the Burmese border, where he hoped to stay since he could not stay in Burma or Sri-Lanka. After China joined the war against Germany, he was interned in China and was repatriated to Germany in 1919.

World War II

In 1939, with the British declaration of war against Nazi Germany, Nyanatiloka and other German-born Sri Lankans were again interned, first again at Diyatalawa in Sri Lanka and then in India (1941) at the large internment camp at Dehra Dun.

Nyanaponika Thera

Ven. Nyanaponika Thera (Siegmund Feniger) was born in Hanau, Germany on July 21st, 1901 as Siegmund Feniger, the only child of a Jewish family. In 1922, he moved with his parents to Berlin, where he met with other German Buddhists and also had access to Buddhist literature in German language. He first came across the writings of Ven. Nyanatiloka Thera, the former German violin virtuoso Anton Gueth, which had already been published in Germany. Young Siegmund had learned Ven. Nyanatiloka Thera had established a monastery for Western Monks on an island lagoon (opposite the Railway station) Polgasduwa, Dodanduwa named Island Hermitage.

This news stirred his conscience to come to Asia and become a Buddhist Monk. However this did not materialize for some time. In 1932, his father died and he did not want to leave his widowed mother in the lurch.

Then Adolf Hitler came to power in Germany.

In 1939, after the Nazis invaded Poland, Ven. Nyanaponika Thera arranged for his mother and other relatives to come over to Sri Lanka. Through the influence of her son and the generous hosts she embraced Buddha Dhamma and became a devoted Buddhist. She died in Colombo in 1956.

When the Second World War broke out in 1939, and the British Government had all German males resident in their colonies consigned for internment suspecting them to be German spies. The internment was first at Diyatalawa Army cantonment in Sri Lanka and later at Dehra Dun in northern India.

Despite these traumatic experiences as prisoner of war, during this period, Ven. Nyanaponika Thera completed the German translations of the Sutta Nipata, the Dhammasangani (the first book of the Abhidhamma Pitaka) and its commentary.

He also compiled an anthology of texts on Satipatthana Meditation. This was commenced at Diyatalawa, and the rest of the writing completed at Dehra Dun, India.

"The Heart of Buddhist Meditation: The Buddha's Way of Mindfulness" is one of the finest works of this monk.

With the cessation of war, the two monks were released from internment at Dehra Dun and returned to Sri Lanka in 1946 and resided at the Island Hermitage, Dodanduwa. In early 1951 they were conferred citizenship of Sri-Lanka.

In 1946, Ven. Nyanatiloka Thera was offered a hermitage in the Udawattekelle Forest Reserve, and being advanced in age preferred the cooler climate of Kandy rather than the hot and stuffy sea - coast climate of Dodanduwa.

In 1947, Ven. Nyanaponika Thera too joined him at the new Kandy Hermitage.

In 1952, both Venerable Nyanatiloka Thera and Nyanaponika Thera were invited by the Burmese (Myanmar) Government for consultation in preparation of the Sixth Buddhist Council, to be convened in 1954 to re-edit and reprint the entire Pali Canon and its commentaries. On the conclusion of the consultations Ven. Nyanaponika Thera stayed in Burma for a period of training in Insight Meditation (Vipassana) under the renowned meditation teacher Ven. Mahasi Sayadaw Thera.

The experience he gathered motivated him to write his best known work, "The Heart of Buddhist Meditation" published by Buddhist Publication Society with many editions and translated into more than seven languages. This is a prescribed text in universities in the Study of Buddhist Meditation.

In 1954, the teacher and the pupil returned to Burma for the opening ceremonies of the Council, which was held in a cave-structure built similar to the Sattaprani Caves in Rajagaha (Rajgir) of India, where first Buddhist Council was held. For the closing ceremonies in 1956 Ven. Nyanaponika Thera went to Burma alone as his teacher was indisposed.

In 1957, the health of Ven. Nyanatiloka Thera deteriorated and he moved to Colombo for medical attention. Finally on May 28th, 1957, the great pioneering scholar monk passed away and was accorded a State Funeral at the Independence Square, Colombo.

His ashes were enshrined at the Polgasduwa Island Hermitage, Dodanduwa and a tombstone was built in his memory. Ven. Nyanaponika Thera, thereafter keeping up to the request of his teacher, revised Ven. Nyanatiloka Thera's German translation of the complete Anguttara Nikaya, retyping the five volumes in full by himself, and also compiling a forty paged index to the work.

Six months after the death of his teacher, the career of Ven. Nyanaponika Thera was to be launched in a new direction, a permanent contribution to the spread of Buddhism worldwide. A prominent lawyer in Kandy A.S. Karunaratne suggested to his friend, Trinity College teacher in retirement Richard Abeysekera, that they start a society for the publication of Buddhist literature in English, mainly to be distributed abroad. The unanimous decision was Ven. Nyanaponika Thera in the Udawattekelle Forest Reserve Aramaya would be the best director of the Institution.

Thus on the New year's Day of 1958, the Buddhist Publication Society (BPS) was born.

Devoting his entire time and energy to writings of the society, encouraged others to write, collated important Suttas, translated them and had them published. In addition to his own writings, he had 200 Wheel titles and 100 Bodhi Leaves (booklets) authored by numerous scholars. They were published at the BPS after rigorous scrutiny and his able editorship.

With advancing age having a heavy toll on his strength, in 1984, Ven. Nyanaponika Thera retired as editor of BPS and in 1988 he retired as President, accepting appointments as BPS's distinguished Patron. His fame and recognition as an exponent of authentic Theravada Buddhism is unrivaled.

The Dhamma message has now reached all corners of the world.

In 1978, the German Oriental Society appointed him an honorary member in recognition of his combination of objective scholarship with religious practice as a Buddhist Monk. In 1987, the Buddhist and Pali University of Sri Lanka at its first convocation, conferred on him its first ever Honoris Causa Degree of Doctor of Literature.

In 1990, he received the Honoris Causa Degree of Doctor of Letters from the University of Peradeniya. In 1993, The Amarapura Maha Sangha Sabha to which he belonged for 56 years, conferred on him the honorary title of Amarapura Maha Mahopadhyaya Sasana Sobhana (The Great Mentor of the Amarapura Maha Sasana Sabha, Ornament of Teaching).

His last birthday which fell on July 21st, 1994, was celebrated by his friends and the BPS staff with the release of the BPS edition of his book "The Vision of Dhamma", a collection of his writings from Wheel and Bodhi leaves series. On the 19th of October 1994, the last day of his 58th Rains Retreat as a Bhikkhu, he passed away in the quietude of the Udawattekelle forest hermitage.

Chapter 03

Blind Logics and Logistics of Modern Society

I believe the modern society lacks perspective in life.

Good example is America.

If you look at the presidential campaign, I am amazed why America cannot find better alternatives to Trump or Hillary.

Both are out of touch with reality and living in their own worlds of misrepresentation and misconceptions.

There are only two parties.

There is no place for a third party.

There is no avenue or emergence of alternative views. Both parties are bound up with hidden pressure groups and money. The pressure groups with ulterior motives, do the propaganda.

The average citizen is led down the garden path of no return. He or she does not have any space left to mingle with.

I do not know where the FBI and NSC/NSA stand, except for eavesdropping globally on every human being born or not born yet.

England where the "Mother of Democracy" supposed to have originated is getting too much direction from bureaucrats from Brussels and British people had no alternative but to say firm, No.

But the three main parties do not know how to get out the mess they are in. How they are going to lead the world in modern politics, is a blind guess of any docile polity.

At least United Kingdom has an alternative party.

But it is all bickering within and outside the individual parties that is evident, not concrete policies.

Taking up the leadership mantle, is the only goal, of the top brass.

Nothing else is evident?

Come to Ceylon, it is all about killing the democracy and the opponents at all costs. It opened a big hole for the descendants of a single family to rule. A modern dynasty in evolution.

We are unable to get rid of the monster called the elected president. Election to this post is devious by any stretch of power politics.

We have a vestige of alternative media and most of the others are run by undesirable agents. It is believed that some of the media have connections to drug cartels or warlords.

Now with three women leading the world, we the men folks can say good bye to politics and start polishing the remaining spirits left in our bottles or the drums. I can assure you that these three women will not leave the world in a better footing than when they took over the control.

I guess it would be worse.

The simple analysis above is not at all for political reasons but to show the logistics part of governance. I used politics as a preamble to illustrate the bigger picture in our social evolution.

The main theme however, is to address, how we destroy the creativity and freedom of inquiry from the cradle.

There is an outrageous attempt at misinformation from birth to teen to coffin.

One bizarre observation is all too, apparent.

The right to be born free without prejudice is no more.

In this part of the world the female infant (including China) is unwanted commodity and if possible aborted before birth by default.

If they are born alive they go through hell.

In our country the infant has to be born before December 31st, to beat the mad rush, for entry into a National School. According to doctored documents, most of them are born within a single house bordering a National School by a distance of one mile. The statistics emerging from houses within a mile radius is staggering.

The highest population density by birth in the entire world!

The second observation is also curious.

Once born these kids have no right or choice.

There is only one goal.

They have to get the highest grades, in the year five examination. The right to free education of choice of the child is negated at the (exit) door of the labour room. Child is sent to the preschool very early to learn a foreign language before he/she has become verse with the mother tongue. Do not get me wrong here, second and perhaps third language are vital to move upwards on the ladder of education, in this country and elsewhere in the modern world.

The third observation is global in nature but has a distinct local aberration.

With proper education one moves up the ladder fast and the education is the only way forward. It is a myth perpetuated as far as our country is concerned. In our parliament the richest ones have not even gone to the fifth standard.

Their school is local mafiosi and truancy is the gateway froward.

The fourth observation is the most common or the default form.

The failures of the year five examination are made to believe, there is a way around the education obstacle. If they fail in the normal mode of schooling, they should attend the private tuition classes.

It is a booming industry for pseudo-educationalists.

Fifth a bigger conspiracy is the free books donation to all students at primary level. I have written enough on this, suffice to summarize the cardinal points.

These books are outdated by 30 years and no effort is taken to revise and update them. The bigger picture is, having donated books free, the education department washes its hands off on education and does precise little to improve the quality of education and use it as a platform to provide employment to untrained political stooges. My real antagonism to this program is on a different ground and reason.

It shuts off all creative writers.

The student never look at alternative books or literature. Books are dumped in Tamils and Sinhala languages and practically never in English.

I thank god I was born before this era.

The sixth is the biggest conspiracy of all.

Changing the language of instruction from English to Tamil and Sinhala is the Biggest Conspiracy. This was done without research and on putative reasoning. The English language (Schools under British) was penetrating even the remotest villages in this country except a few under the British. Our so called "Freedom Fighters" under the British did see the impact and the ominous sign. If English was the instruction language, the racial and caste barriers would be wiped out in a shorter time.

May be one or two generations.

The cunning politicians to hoodwink the masses, held onto this Swabaha (India did not with Insight) Program, the the end result was war, proliferation of caste system and anarchy.

I strongly believe that if British remained for another 20 years, this sinister ploy of the local politicians would have been curtailed to a manageable proportion.

The biggest loss was books in English especially in science and technology. In computer science, one cannot write program language in Tamil or Sinhala.

The code was fixed to the English Keyboard.

Let me cut short, some of them by shear chance, get to the university. My general estimate is, by that time, they have lost all their creativity but are loaded with outdated (by about 30 years) book knowledge in Swabasha.

Never read an English (French/German) book except those who did English as a subject.

The bottom line is that the parents, teachers (most are not trained in education psychology) and specially the religious teachers of all faiths (they are not taught that the religion is a blind faith, an opium, that hinder creativity and advancement in scientific knowledge) put all sort of rigid dogmas in their minds from cradle to teen to coffin.

By the time, they are teens we have created JVP activists, LTTE activists and currently religious activists including ISIS and BBS.

I often have two very simple question for any inquiring mind.

1. What have you gain/ed from your religion (except perhaps extremism)?

If one says that I have become a good (not right) person, the next question I pose is;

2. Do you really need a religion to become a good or bad person?

Isn't it common sense?

They are flabbergasted and do not ponder a philosophical or scientific question beyond that point. Neither Karl Marx nor Einstein got any inspiration from previous dogmas (religion included) but their contributions to the society and mankind in general (particularly to the thinking man) are enormous.

Chapter 04

Why Philosophy?

Philosophy is a first order discipline. All others are second order disciplines that include, science, politics and religion. The original idea was to expose the conspiracies in philosophy in this book. But on second thought, I decided to leave ancient philosophers and their dogmas untainted.

Of course the religion and politics carried forward the conspiracies of those holding power (according to Sir Bertrand Russell). Not only philosophy, even the science and medicine did not flourish in subsequent centuries after the demise of these scholars.

It was in the middle of the 20th century that science and medicine started to emerge from the dark ages.

Amidst these changes, unfortunately, American scientists joined the band wagon of conspiracy.

Hippocrates

Hippocrates of Kos (Greek. 460 BC – 370 BC), also known as Hippocrates II, was a Greek physician of the Age of Pericles (Classical Greece), and is considered one of the most outstanding figures in the history of medicine. He is referred to as the "Father of Western Medicine" in recognition of his lasting contributions to the field as the founder of the Hippocratic School of Medicine. This intellectual school revolutionized medicine in ancient Greece, establishing it as a discipline distinct from other fields with which it had traditionally been associated (theology and philosophy) with, thus establishing medicine as a profession.

However, the achievements of the writers of the Corpus, the practitioners of Hippocratic medicine, and the actions of Hippocrates himself were often commingled; thus very little is known about what Hippocrates actually thought, wrote, and did. Hippocrates is commonly portrayed as the paragon of the ancient physician, and credited with coining the Hippocratic Oath, still relevant and in use today.

He was also credited to be the first person to clearly state that diseases were caused by natural reasons, and not by superstition and wraths of god – a belief that used to be widespread in the ancient ages.

He is also credited with greatly advancing the systematic study of clinical medicine, summing up the medical knowledge of previous schools, and prescribing practices for physicians through the Hippocratic Corpus.

Plato

One of the foremost influential figures in the ancient philosophy, Plato was born somewhere around 428 BC – 423 BC in the ancient Athens. (428/427 or 424/423 – 348/347 BC). A bona fide student of Socrates, another philosophical legend from ancient Greece, he was named Aristocles by birth, but later earned the nickname of Platon (meaning broad) courtesy of his broad built. Amid the political crisis, most noticeably the execution of his teacher Socrates, Plato needed no more conviction to leave the virtues of Athenian politics.

He the founder of the Academy in Athens, the first institution of higher learning in the Western world. He is widely considered the most pivotal figure in the development of philosophy, especially the Western tradition. Unlike nearly all of his philosophical contemporaries, Plato's entire collection of writings, are believed to have survived intact for over 2,400 years.

Along with his teacher, Socrates, and his most famous student, Aristotle, Plato laid the very foundations of Western philosophy and science.

Alfred North Whitehead once noted: "the safest general characterization of the European philosophical tradition is that it consists of a series of footnotes to Plato."

In addition to being a foundational figure for Western science, philosophy, and mathematics, Plato has also often been cited as one of the founders of Western religion and spirituality.

Friedrich Nietzsche, amongst other scholars, called Christianity, "Platonism for the people."

Plato's influence on Christian thought is often thought to be mediated by his major influence on Saint Augustine of Hippo, one of the most important philosophers and theologians in the history of Christianity.

Plato was the innovator of the written dialogue and dialectic forms in philosophy, which originate with him. Plato appears to have been the founder of Western political philosophy, with his Republic, and Laws among other dialogues, providing some of the earliest extant treatments of political questions from a philosophical perspective.

Plato's own most decisive philosophical influences are usually thought to have been Socrates, Parmenides, Heraclitus and Pythagoras, although few of his predecessors' works remain extant and much of what we know about these figures today derives from Plato himself.

The Stanford Encyclopedia of Philosophy describes Plato as "One of the most dazzling writers in the Western literary tradition and one of the most penetrating, wide-ranging, and influential authors in the history of philosophy. He was not the first thinker or writer to whom the word "philosopher" should be applied. But he was conscious about how philosophy should be conceived, and what its scope and ambitions properly were. By this conviction, he transformed the intellectual currents with which he grappled with. The subject of philosophy became, a rigorous and systematic examination of ethical, political, metaphysical, and epistemological issues, armed with a distinctive method. The strict methodological approach was his invention.

Few other authors in the history of Western philosophy approximate him in depth and range: perhaps only Aristotle (who studied with him), Aquinas and Kant would be generally agreed to be of the same rank.

Chapter 05

Aristotle

Aristotle was a Greek philosopher and a scientist born in the city of Stagira, Chalkidice, on the northern periphery of Classical Greece. His father, Nicomachus, died when Aristotle was a child, thereafter Proxenus of Atarneus became his guardian.

At eighteen, he joined Plato's Academy in Athens and remained there until the age of thirty seven (347 BC). His writings cover many subjects – including physics, biology, zoology, metaphysics, logic, ethics, aesthetics, poetry, theater, music, rhetoric, linguistics, politics and government – and constitute the first comprehensive system of Western philosophy. Shortly after Plato died, Aristotle left Athens and, at the request of Philip of Macedon, tutored Alexander the Great starting from 343 BC.

According to the Encyclopædia Britannica, "Aristotle was the first genuine scientist in history and every scientist is in his debt."

Teaching Alexander the Great gave Aristotle many opportunities and an abundance of supplies. He established a library in the Lyceum which aided in the production of many of his hundreds of books. The fact that Aristotle was a pupil of Plato contributed to his former views of Platonism, but, following Plato's death, Aristotle immersed himself in empirical studies and shifted from Platonism to empiricism.

He believed all peoples' concepts and all of their knowledge was ultimately based on perception. Aristotle's views on natural sciences represent the groundwork underlying many of his works.

Aristotle's views on physical science profoundly shaped medieval scholarship. Their influence extended into the Renaissance and were not replaced systematically until the Enlightenment and theories such as classical mechanics. Some of Aristotle's zoological observations, such as on the hectocotyl (reproductive) arm of the octopus, were not confirmed or refuted until the 19th century. His works contain the earliest known formal study of logic, which was incorporated in the late 19th century into modern formal logic.

In metaphysics, Aristotelianism profoundly influenced Judeo-Islamic philosophical and theological thought during the Middle Ages and continues to influence Christian theology, especially the scholastic tradition of the Catholic Church. Aristotle was well known among medieval Muslim intellectuals and revered as "The First Teacher". His ethics, though always influential, gained renewed interest with the modern advent of virtue ethics. All aspects of Aristotle's philosophy continue to be the object of active academic study today. Though, Aristotle wrote many elegant treatises and dialogues – Cicero described his literary style as "a river of gold" it is thought that only around a third of his original output has survived. All these philosophers could not avoid religion as an exegesis and that retarded the progress of science.

Chapter 06

Religion

Why the decadence?

We have gone from basic good and bad moral concepts to to total decadence.

I have very a simple answer.

It is the selfishness of the majority.

Nobody stands up for justice.

This is a malady that comes after 3 decades of war and killing. As long as one who was killed is not in one's particular power group (ethnicity), an average citizen looks the other way, adoring the killing machines.

The life is only a private affair and some spend millions for self propagation. A sum of over 10 million spent on a politician's leg. The rest should go to the government hospital and wait in the queue.

What contribution the religion has made?

We are supposed to have four of the best religions in this country, yet we had war over 30 years and five years after ending the bloody war nobody speaks about reconciliation.

War in politics is marketable but peace has no market value for politicians.

But the average citizen loves peace.

The religion does not contribute to peace bur war, Middle East in crisis, is good example.

It may filter down here.

Is it better without a religion?

For want of better ideas, my philosophical tenet is that the world in historical sense (Bible War) and for the current sustenance, the contribution of the religion is marginal at its best.

A country without a religion is where I want to be.

I would be happier then without a dogma in my head.

I cannot wait till I go to heaven after death to seek happiness.

I want my happiness now.

I want to live in happiness today, I can't wait for,

"a tomorrow", that has not yet come!

Chapter 07

Meaning of Freedom

The dangling of the carrot is the proverbial statement for an offering that is apparently "feasible but not tenable", that is presented to hoodwink the unsuspecting masses.

Now the operative word is the FUD, a shorthand for Fear, Uncertainty and the Doubt.

FUD the triple of Fear, Uncertainty and the Doubt that is created in the minds of masses by corporate giants and is utilized to sideline the competitors.

Good example is the operation of Microsoft against the competitive Linux distributions. When Linux was emerging as an operating system the FUD was amply utilized to distance the would be computer enthusiast and lure him / her to Microsoft Windows.

But being intellectuals most of the Linux enthusiast overcame this by sheer courage and community spirits. Linux was very flexible and now quite unintentionally FUD is working against its originator. Community spirit prevailed in the Linux community and the Free Software Foundation laid the foundation in overcoming the FUD syndrome. The spirit of Free Software Foundation / Linux is the gift of original software to the Internet Community.

As long as the authorship (not ownership) is quoted any modification could be done to the original copy to improve it and remove bugs. However the improved version has to be re-released back to the pool for further development by the community.

It is called the copy left and not copy right.

So the natural evolution of the software takes leaps and bounds within a short space of time for the betterment of the masses.

Much of this principle is very much close to the Buddhist principle of "Dhamma Dhana" but with a difference. What is given away is recycled with added improvement in quality and it is given away for another cycle like the "Sansara" cycle never ending.

One need not improve "The Dhamma" but only have to understand its underlying principles and spread the message of goodwill.

It is interesting to note that a Buddhist monk who had appeared in a court to protect his Pirith cassette, he had been marketing.

Is the "Sabbha Dhanam Dhamma Dhanam Jinathi" spirit of Buddhism being betrayed?

This monk should read the evolution of Linux Community and the General Public License that came with the Free Software Foundation. This Buddhist monk does not understand that the Gift of Dhamma is the supreme gift and it should not be commercialized. A blank CD / cassette is only Rs.25 / 50 and anybody who is selling a Pirith CD / cassette for more than Rs.50 is vandalizing these principles.

If Buddha is alive today, I wonder what he would utter.

Would be an another occasion for "Udana Whakya" similar to "Sabbha Dhanam Linux Dhanam Jinathi" and would have a little smile at the corner of his mouth. There would be few more verses added illustrating the illusion of freedom and liberty of the mind filled with lust, hate and delusion.

Are we truly free and liberated?

It is something worth pondering.

All Sri-Lankans whatever the creed, class or caste, may be, believe that they are prisoners of their own conscience.

Sinhalese believe that they have to be free from interference by Tamils.

Tamils believe they are persecuted by the majority.

Are the Tamils free from persecution by their masters?.

The Muslims, believe they have to be free from indulgence from both Tamils and Sinhalese.

Million dollar question is, could the piece mongers (foreign ambassadors of dubious credentials) allay the FUD syndrome?

They in fact created it.

We are in this vicious cycle of suspicion, the LTTE, JVP, party in power and party aspiring to come to power are spreading the FUD syndrome viciously to its maximum zenith (peak). What is needed is to have FUD syndrome at its nadir (depths).

The international community spreads the FUD syndrome to help the propagation of their corporate agenda with dangling the proverbial carrot.

What is happening in the political field in Sri-Lanka is the insensible manifestation of FUD syndrome.

At a philosophical level what Socrates said to an aggressive follower who would stretch his arm for a punch is relevant even today.

"Your freedom ends where my nose begins".

The international community should realize that they are very close to every one's nose.

It is nice to recapture what Nazurudeen stated about his failure to marry.

It is said that he had been to all the countries looking for a girl he liked. Ultimately he found one but the girl who matched his interests was also looking for man whom she wanted but alas for him / her that man was not Nazurudeen.

We have being looking for a suitable match maker for Sri-Lankan bride. Now we are left with a whole harem of foreign ambassadorial girls (donating their vices).

At what price we do not know?

This is where one has to ponder and ask the vital question what is the price of freedom and what is the meaning of liberty?

Poor man on the street does not understand the jargon he will never be freed from his poverty.

As long as there is corporate mind set in making profits without sharing the wealth of information with its competitors to improve the quality and availability at an affordable price the gap between the rich and poor would widen.

Free market philosophy is not for liberating the masses.

They are there to increase the profits.

The belief that profits would filter to the masses is a grand myth.

There is no community involvement as seen in the Linux community. Linux community feels that they are liberated from the corporate giants. Until and unless we are free from both local and foreign corporate giants who are hell bent on making profits and exercise their monopoly, the freedom of choice is an illusion to the majority.

Only the minority will have the freedom to enjoy and exploit.

This corporate mind set should change to community mind set and community involvement.

There is a limit that the free market philosophy can stretch but beyond that point there is diminishing returns.

Like in the Linux community somewhat similar orientation has to take place in the business community in Sri-Lanka and worldwide for the true meaning of freedom to be enjoyed by all.

Not a privileged and selected minority.

Freedom that costs is meaningless.

Selfish gains but no devotion as is preached by all the major religions is not relevant to the corporate mind set.

Microsoft agents doing a few community projects (to lure a few deprived clients in the periphery) without shedding their corporate mentality is of little benefit to the masses.

A wholesome benefits want accrue.

If we do not think in radical terms and initiate changes now, there going to be massive uprising, hitherto unknown in the past. Some of the manifestations in the world today are a sign of this frustration building.

Not only with the poor but in the middle class too.

The middle class gave the stability but it is fast disappearing.

With the wide use of the Internet, we would hear of more and more of evolving crisis.

People would like to see the credibility in their leaders which is sadly lacking in the world because of corporate giants pulling their strings behind their leaders.

They determine policy not the larger masses.

The masses would ask what is the meaning of this freedom?

That is the starting point of the crisis, in any modern society / country in this century.

Leaders have to be groomed with social values and ideals and not individual profiles and party profiles. Leaders should not be trained only to run the corporate giants. Leaders should be trained to lead the community they live.

Emulating western values only, going to be, not enough and begging bowl mentality should give way to equal partners in international dealing whether the country is big or small.

Some of the eastern values of sharing and caring should take precedence over profits. Then only one can call the citizens of sovereign nations are free and liberated.

United Nations will fail in their duty, if they have only Human Right Charter for cosmetic exercises. In that case the amount money that is spent on UNO could be better utilized for some other ventures.

They should rewrite these Charters.

Profit beyond certain acceptable levels should be prohibited or a certain level of compulsion to do community research both eco-friendly and community-friendly should be encouraged.

What is the big idea of having a few rich people and million and millions of poor souls?

How can a man like Tyson a champion one day and then a pauper the next day (in his twilight years)?

It is not acceptable in USA and for that matter any other country in this world.

Where is the social security?

England had a very well organized social security system and in another 50 years time it is going to be all private pensions. These questions that are raised in the West and are equally relevant to us. The people who raise these voices have no industrial or corporate muscle.

Party politics seem to have ruined the entire world. Most of the party leaders worldwide are gullible liars and some of them are of course pathological in nature.

Can we trust these leaders?

This is what is emerging in the West.

The covert political mechanisms are firmly embedded and established, in the current system, only the corrupt and the rich can rise to the top.

So, talking about freedom whether it is in the West or East is futile to the average man on the street. They just get the kick out of kicking a party out of power to get another miserable party into power and languish till the next opportunity, to make the same mistake again.

In reality, is it a Gamble or a Casino?

One day cricket was much more interesting until this Casino bug bit it in full.

At the end of the day, all the so called democratic exercises are futile and the freedom of choice, a bad dream and a nightmare, just as well, to be forgotten by the majority.

The liberation that the average man is looking for in economic and social fields is not achieved.

Goals are set but never achieved due to lack of penetrating insight and vision.

What is done is patchy and ad hoc.

We are blindly going through the cycle of events till events take control of our freedom.

Are we truly liberated?

The answer is firm, No.

Freedom

Four freedoms elegantly expressed by former US President F.D.Roosevelt have given way to FUD factor. The freedom of speech and expression (state controlled media and private media with vested interests), freedom of worship (desecrating religions), freedom from want (poverty) and last but not least freedom from fear (terrorism) has no meaning today worldwide.

The last factor the fear (from both physical and psychological) is utilized to gain undesirable motives.

The psychological fear is the deadliest of all and our doctors are using this to the detriment of the profession (both public and private). "Fear of Death" is used as a mean to market pseudo rituals, putative medicines and divine interventions.

If you look at the Roosevelt's statement and American mentality, there is a big void there.

There is no place for a guy/girl who wants, not to be bonded or branded by a religion or a dogma.

There is no place for an atheist.

One has to born or bound to a religion.

Late Bertrand Russel had this problem.

Freedom Revisited

I just did a survey of our personal and political freedom from the time of independence and suddenly discovered seven (7) stages.

This can be applied to an elephant in captivity but in the case of the elephant the poor animal has few stages and no political freedom at all.

The elephant can be domesticated at whims and fancies of the human.

I will list them first and the reader can have his or her own classification.

1. Stage One - **Born Free**

That is the stage we enjoyed for over a decade. Then we went in search of the second stage since the freedom did not filter down to everybody.

2. Stage Two - **Freedom Struggle**

We lacked economic freedom at this stage.

3. Stage Three- **Freedom Fighter** (Wasted time in our history)

Never got the freedom we were fighting for.

JVP and LTTE are solid examples.

4. Stage Four - **Freedom left to be Abused**

This is the stage where our politicians (all shades), in the exercise of their political power abused our freedom left, right and center (they are the only ones who gets a pension after five years in parliament for doing sweet nothing).

5. Stage 5 -**Freedom Lost**

This is the stage we lost all our freedom in the name of new constitution. Even the judges lost their freedom to political power and their whims and fancies.

6. Stage 6 - **Freedom to be abused by those who are in Power.**

This is the most interesting phase where anybody who is somebody with some political power (this includes government officers) can become corrupt with impunity with no legal barriers.

7. Last or the 7th stage is **"Freeing the Freedom"** from clutches of Power.

Unfortunately we have not reached this stage since there are enough legal barriers like 2/3 (two third) majority and referendums., etc. We have to go behind a Bo Tree and Pray for Divine Intervention. Unfortunately foreigners including Indians and Westerners can do sweet nothing since we are a Sovereign State and one has to leave sovereign bars (Gold) in front of the altar of Demo=Crazy.

Chapter 08

Political Wisdom

This is an attempt to explore "political science" and preferably its wisdom and bare its bones free from party bondage. I am not an expert in politics but because somebody has aligned science as a prefix to make it more attractive to the ordinary, I have the right to dissect it in current context.

From Plato's Republic to the present, what emerges is the conventional wisdom of **"the right view"** is the view of those who are currently **"holding power"**. This dissection would not come favorable to any G.O.P (Grand Old Party) not in power, whether, republican, democratic or socialist.

First of all, I dug into the definition of political science, true to its credentials, I have not found a worthwhile definition. In any case, I found that the term scientific cannot be applied to politics.

When Socrates philosophical dilemmas were terminated by death (execution), let alone science even philosophy was not accommodated by the rulers. When Jesus was crucified, the first anomaly of democracy was born.

If politics is to silence the contrary view there is no semblance of science in its inception.

The appendage "science" was a clever instrument by which rulers, the powerful ones, found a way to dub the innocent masses.

The westerners' obsession to democracy is simple and cunning. It gives them a subtle tool to insinuate their ideas to alien cultures. The fact of the matter is that mankind has not found a foolproof method to govern itself in prosperity or in adversity.

What is current and vogue takes precedence to prudence in politics. Invariably politic is not scientific in its origin or its evolution.

My stress in this short provocation is simple.

How can one arrive at the term of office in years?

Isn't it arbitrary?

It varies from three to four to five to six years to almost infinity for some dictators.

Since, I am using a scientific argument, the term of office should be related to the biological life of the voter.

Assuming one's average productive lifespan as sixty years when most of us retire to insignificance except politicians, my observations are as follows.

The results of the Genome Project are producing some interesting findings.

I can summarize few of them for the average voter.

1. Man has only 50 cell cycles to live.

2. If parents are old the cell cycles are reduced proportionately.

3. Most of the cell cycles are spent in uterus (before one is born).

4. After puberty one is left with about 6 to 8 cell cycles, each cycle roughly eight to twelve years in span.

5. In some, the duration of a cell cycle may be as short as five years.

6. Man starts aging at 30 years (not 35 as was believed).

7. Assuming one is eligible for voting at 20 years and the parents are responsible for looking after them and educating them up till then, any government has to manage the voters 10 years before they start to age and 30 years thereafter (in a 60 year life span), roughly 40 years in total.

8. If this is divided by the individual cell cycle duration, one is left with 5 voting cycles for every man / women.

9. If rulers are sensible the voters should not be abused more than this number of voting fiascoes, in their life span.

10. So my scientific reason and of course a biological one is that elections should be held once in eight years.

Better still once in 10 years.

11. The advantage for the rulers is that the voter has gone one full cell cycle (in fact older and more mature by age) before the next election. They invariably have forgotten all the promises given by the elected, so re-election is not going to be dicey.

12. I believe that eight years is more than enough for the elected to swindle enough money and get rich for a generation (before they get old). Even, if one loses the next election that would be offset by the long duration in power (occupation of the hot seat) whether it is in America or Russia or Sri-Lanka or anywhere in this world.

Reflection of our Polity

Reflection of the thirty year cycle (equivalent to the period of growth and maturity in a biological cycle) of politics in Sri-Lanka from 1969 to 1999 is illuminating.

I pick this period because all the politics that started from, 1969 (both parties in power and opposition) was an aberration of immense proportion, in our history.

When we were in the campus we thought that the youths were oozing with enthusiasm and trust.

We all wanted governance without corruption but clean.

What did we get?

Absolute corruption.

Soon, I realized even before the first appointment in Government Service was due, the underhand dealings were made by some to get attractive posts. It was not bad as today but it was evident by example. Before we reached our maturity (30 years- before start aging) we realized that the ideal world was yet to become.

Most of us, by default, left our shores for greener pastures, disillusioned by the apathy in Health Sector, Western countries fared only a little better, the discrimination was evident by practice.

But by 1977 people voted for more and more corruption.

Corruption and get rich quickly was the motive of politics.

Entry of terrorism and its appendages here and abroad laid the firm foundation for cruelty plus corruption.

There was no turning back.

I would take University Structure to illustrate how it deteriorated to present status. By 1969, then Minister of Education who rode rough with (his son was in the university) university students and developed an aversion, decided to send an army platoon to the university gymnasium for preparation for the Independence Day celebration.

Incident that followed was the forerunner for students getting involved in 1970 elections. Prior to this except perhaps Arts Faculty, all the Student Unions were run by independent minded students and generally politics was bane of student community.

By 1973, university did not have a proper Vice Chancellor, but a Competent Authority and politics was eroding the students' freedom and identity. Bursary was withheld and gradually university administration became politicized. Convocations were not held and the administration allowed the student discipline to deteriorate rapidly, especially as regard to the ragging.

General Election was postponed to hold Non Aligned Summit and all the ingredients for dissent were accumulating.

By 1978 with the revision of the University Act (to suit political agenda), the hitherto, independent university lost its integrity as an autonomous institution. Student intakes were determined by Education Ministry not by the University academics.

Practical examination for science was stopped on an arbitrary issue.

Vice Chancellors were appointed on political favoritism and not on university standards which is a universal practice today.

It is true that, lot of students go through difficulties including financial and language abilities. Unions are dictated by factions that make student life more miserable than it should be.

No free discussions, no independent unions and no sense of direction. Arbitrary decision making is the norm from the top to the bottom.

In other words, in little over 30 years, the whole structure has cracked from within and without.

By the time an individual reaches maturity (Prime Age of 30 years), by strange coincidence, the structure that was established by the British and was maturing until mid sixties was replaced with an arbitrary change of political wisdom.

The new form of administration that replaced the old failed to establish accountability in the university system, leave alone the academic standards.

There was a desperate attempt to undermine the university system.

Where is the political wisdom when it comes to the most important Higher Education?

Coming back to my interpretation of political life cycle (30 years to maturity and aging beyond that point and 5 cycles of 8-12 year duration) of human beings, what we aspired as students are no more, a new anarchic and frustrated structure is developing.

What has happened to the autonomous university of yesteryear is happening to the body politics of Sri-Lanka, vigorously and alarmingly.

In 30 years (full maturity of a human being) we have turned everything upside down. Dreams of yesteryear generation of students were shattered and a new wave of youth unrest is in the offing.

By the time they reach maturity, in 8 year cycles or become old adults they are entrapped in corruption. Then in the last few cycles of their life they start ruminating what went wrong without any inclination to rectify the defects.

There is no discussion what ought to be done.

Patch work is done on a regular basis with each change of administration at the top.

The cycle of corruption begins with a cycle of anti-corruption. The corruption always wins heads down, whether it is in this country or abroad.

Politics has done nothing in the long run.

That is my scientific evaluation.

One is free to disagree or dissect it again in correct perspective.

In the mean time, population crisis, health crisis, environmental crisis, global warming, energy crisis, land crisis and human crisis (ethnic and religious) would aggravate and take us to the next thirty year cycle. Last thirty years of last century and the next thirty years of the present century (which is a generation cycle) would determine how prepared are we for a major crisis.

It is impending.

The system has failed.

The education have failed.

The university system, one believes, where the brains of the last century were educated was unable (not fit for purpose) to produce leaders with leadership qualities to take problems, head on.

Where will the leaders come from?

They have come from the local mafiosi.

Where and what we are heading for is immaterial.

We are heading aimlessly for catastrophe.

It is a gloomy scenario, unless of course, we look at, in depth and take necessary steps right now.

Chapter 09

Basic Tenet of Human Civilization

I think the basic tenet of all civilizations, whether our own or galactic won't change.

I fear to think but the underline theme is paranoia.

The outcome of paranoia is, to dominate first and destroy next. From Buddhist Civilization (Avihinsa has no dominant role, just think of Dalai Lama with American and Indian support) to Muslim Civilization to Judo-Christian, Western Civilization to current Dollar Civilization we engineered, lot of software and hardware for our civilization.

It will invariably yield to Chinese Yuan.

Nobody talks about Yen now!

Yuan rules and Dalai Lama succumbs.

What will become of us after Yuan (becoming a Capitalistic Tool) domination, I haven't got a slightest of a clue or an idea.

I think it is the next global collapse.

Are we ready?

The Premise is that due to blind faith (ISIS is a case in point) or due to greed (America is the golden standard) or insanity (North Korea is a case in point) new civilizations in their forward march destroy themselves in midterm before they are fully evolved or civilized.

We are now in this current wave of destruction.

Galactic Civilizations

Now let me dissect the Galactic Civilizations.

In the galactic scenario, the evolved ones fear the not so evolved ones.

It is a reciprocal relationship.

Evolved ones fear that not so evolved ones would reach their expertise given time and take over the rein.

Not so evolved ones fear the take over by the evolved ones.

The same good old paranoia rules.

Next question is, are there any galactic civilizations without paranoia?

What is their role?

It is a million dollar question.

I have no answer.

If there is any without paranoia, I beg them to visit us now.

We are on the threshold of destruction with Muslim Mafiosi.

Visit by Advanced Civilizations

My alien theory points to the conclusion that they visited us but left convinced that we were not mature enough to interact with them.

I strongly believe they visited us by 1947.

They came to investigate.

About two years after the Atomic Bomb Explosion.

The time interval or the delay is the time to reach us from there origin.

Additionally Area 51 was not evolved to its present level of sophistication, then. It was refurbished after the crash of the alien spaceship (in Roswell) with more advanced technology.

I firmly believe that the alien spaceship was brought down by hostile activity.

By Nigel Calder and John Newell

A very short lifespan for advanced civilizations may be a realistic probability because as our own case shows, technology mushrooms very rapidly and is accompanied by many dangerous effects such as;

1. Overpopulation

2. Depletion of natural resources

3. Pollution and worse of all

4. Rapid escalation of nuclear stockpiles which can easily reach the level where they are capable of wiping out the whole population of a planet.

By the same token, if we were to find significant number of other advanced civilizations in our galaxy it would mean that some at least had been able to overcome the dangers that are now looming over our own civilization.

If they have reached this advanced stage, it would mean that they have been able discard all of their animistic attributes such as materialism, selfishness, territoriality, dominance over others and killing instincts which in a technological civilizations are likely to lead to self destruction.

They would have been able to to mature into higher levels of cosmic intelligence characterized by spirituality, altruism, respect for others, adherence to peace and love for each other.

Joining such a galactic society would indeed be a colossal step forward in the intellectual and ethical evolution of our civilization.

This might also be the reason why we have not yet heard from any such advanced civilizations as may exist.

There must be a galactic rule that says, before a new civilization is invited to join the galactic society, it must show that it is able to overcome the major crisis that probably befall all new technological civilizations.

Since we are still in the midst of this evolutionary problem, there may be advanced civilizations waiting to see how we do before inviting us to join the galactic society of advanced civilizations.

Paul Hellyer, Canadian Minister of Defense

Yet another former government official has made confessions about Aliens and UFOs.

'I was shown inside an alien UFO at Area 51'.

In the last couple of years countless highly ranked officials have come forward with information detailing alien visitations and extraterrestrial spacecrafts.

"At least four (4) known alien species have been visiting Earth for thousands of years" .

A newly surfaced report states that the former government emergency expert, described in detail, the interior of an alien flying saucer inside the tops Secret Area 51 Military Base in the United States.

According to Paul Hellyer, the former Canadian Minister of Defense, a former Canadian Chief of Emergency Measures released astonishing details of alien crafts before his death. Mr. Hellyer, who is 92, revealed the details to a panel of the world's most prestigious UFO investigators.

During the speaking at the "Hearing on ET Disclosure" in Brantford, Ontario, Canada, Mr. Hellyer, said that if he wanted to find out more about the workings of an Alien Space Craft he would ask the 'current chief of emergency measures.'

Not long ago, Paul Hellyer –who was the Canadian Minister of National Defense in the 1960's during the cold war— said in an interview that "At least 4 known alien species have been visiting earth for thousands of years."

Paul Hellyer is the highest ranking person among all G8 countries to speak openly about UFOs and extraterrestrials.

"Decades ago, visitors from other planets warned us about the direction we were heading and offered to help. Instead, some of us interpreted their visits as a threat and decided to shoot first and ask questions after.

It is ironic that the US should be fighting monstrously expensive wars, allegedly to bring democracy to those countries, when it itself can no longer claim to be called a democracy when trillions, and I mean thousands of billions of dollars have been spent on black projects which both congress and the commander in chief have been kept deliberately in the dark."

Mr. Hellyer claims to have seen proof of alien visitations while in office said that: "The reason I know is I interviewed the previous one, who is now deceased, and he went to Langley and the CIA asked if he would like to see one of these crafts.

They flew him to Area 51 and let him go inside one and observe it and make notes and that sort of things.

For years numerous conspiracy theories have revolved around Area 51 which was only confirmed to exist by the government in 2013.

Mr Hellyer added:

"I guess, presumably, it is better to be prepared, to cope with it, if one crashed here".

He was involved in trying to do something positive about it.

"But before he was allowed to go, he had to sign an oath of secrecy and not tell anyone, and during his life he didn't tell anyone including his wife, and an Air Force buddy phoned me, and he was dying of Lou Gehrig's disease and at that point he felt he should tell someone.

"I phoned him, and he gave me a full report of what he saw and the whole idea of the inside of the craft and this sort of thing, and the fact he had been in a brief and many things, but now he felt he could tell somebody, and he thought that would be a good one to tell."

Extraterrestrial Issues

On 3rd June 1967, Hellyer flew in by helicopter to officially inaugurate an unidentified flying object landing pad in St. Paul, Alberta. The town had built it as its Canadian Centennial celebration project, and as a symbol of keeping space free from human warfare.

The sign beside the pad reads:

"The area under the World's First UFO Landing Pad was designated international by the Town of St. Paul as a symbol of our faith that mankind will maintain the outer universe free from national wars and strife. That future travel in space will be safe for all intergalactic beings, all visitors from earth or otherwise are welcome to this territory and to the Town of St. Paul.

In early September 2005, Hellyer made headlines by publicly announcing that he believed in the existence of UFOs. On 25th September 2005, he was an invited speaker at an exopolitics conference in Toronto, where he told the audience that he had seen a UFO one night with his late wife and some friends. He said that, although he had discounted the experience at the time, he had kept an open mind to it. He said that he started taking the issue much more seriously after watching ABC's Peter Jennings' UFO special in February 2005.

Watching Jennings' UFO special prompted Hellyer to read U.S. Army Colonel Philip J. Corso's book "The Day After Roswell", about the Roswell UFO Incident, which had been sitting on his shelf for some time. Hellyer told the Toronto audience that he later spoke to a retired U.S. Air Force general, who confirmed the accuracy of the information in the book.

In November 2005, he accused U.S. President George W. Bush of plotting an "Intergalactic War". The former defense minister told an audience at the University of Toronto:

"The United States military are preparing weapons which could be used against the aliens, and they could get us into an intergalactic war without us ever having any warning.

The Bush Administration has finally agreed to let the military build a forward base on the moon, which will put them in a better position to keep track of the goings and comings of the visitors from space, and to shoot at them, if they so decide."

In 2007, the Ottawa Citizen reported that Hellyer is demanding that world governments disclose alien technology that could be used to solve the problem of climate change:

"I would like to see what (alien) technology there might be that could eliminate the burning of fossil fuels within a generation.

That could be a way to save our planet.

We need to persuade governments to come clean on what they know. Some of us suspect they know quite a lot, and it might be enough to save our planet if applied quickly enough."

In an interview with RT (formerly Russia Today) in 2014, Hellyer said that at least four species of aliens have been visiting Earth for thousands of years, with most of them coming from other star systems, although there are some living on Venus, Mars and Saturn's moon.

According to him, they "don't think we are good stewards of our planet."

The Day After Roswell

The Day After Roswell is an American book about extraterrestrial spacecraft and the Roswell UFO incident. It was written by United States Army Colonel Philip J. Corso, with help from William J. Birnes, and was published as a tell-all memoir by Pocket Books in 1997, a year before Corso's death. The book claims that an extraterrestrial spacecraft crashed near Roswell, New Mexico, in 1947 and was recovered by the United States government who then sought to cover up all evidence of extraterrestrials.

The majority of the book is an account of Colonel Corso's claims that he was assigned to a secret government program that provided some material recovered from crashed spacecraft to private industry to reverse engineer them for corporate use.

Corso was a Special Assistant to Lt General Arthur Trudeau, who headed Army Research and Development, and was in charge of the Foreign Technology Desk. In this position, he would take technological artifacts obtained from Russian, German and other (alien) foreign sources and have American companies reverse engineer that technology. The book contends that several aspects of modern technology such as fiber optics and computer chips were developed by using information taken from the craft.

Philip J. Corso

Philip James Corso (May 22nd, 1915 – July 16th, 1998) was an American Army officer. He served in the United States Army from February 23rd, 1942, to March 1st, 1963, and earned the rank of Lieutenant Colonel. Corso was on the staff of President Eisenhower's National Security Council for four years (1953–1957). In 1961, he became Chief of the Pentagon's Foreign Technology Desk in Army Research and Development, working under Lt. Gen. Arthur Trudeau.

After joining the Army in 1942, Corso served in Army Intelligence in Europe, becoming Chief of the US Counter Intelligence Corps in Rome. In 1945, Corso arranged for the safe passage of 10,000 Jewish World War II refugees out of Rome to the British Mandate of Palestine.

Corso published ***"The Day After Roswell"***, about how he was involved in the research of extraterrestrial technology recovered from the 1947 Roswell UFO Incident.

On July 23rd, 1997, he was a guest on the popular late night radio show, Coast to Coast A.M. with Art Bell where he spoke live about his Roswell story.

In his book "The Day After Roswell" (co-author William J. Birnes) claims he stewarded extraterrestrial artifacts recovered from a crash near Roswell, New Mexico, in 1947. Corso says a covert government group was assembled under the leadership of the first Director of Central Intelligence, Adm. Roscoe H. Hillenkoetter (Majestic 12).

Among its tasks was to collect all information on off-planet technology. The US administration simultaneously discounted the existence of flying saucers in the eyes of the public, Corso says.

According to Corso, the reverse engineering of these artifacts indirectly led to the development of accelerated particle beam devices, fiber optics, lasers, integrated circuit chips and Kevlar material.

In the book, Corso claims the Strategic Defense Initiative (SDI), or "Star Wars", was meant to achieve the destructive capacity of electronic guidance systems in incoming enemy warheads, as well as the disabling of enemy spacecraft, including those of extraterrestrial origin.

General George C. Marshall

George Catlett Marshall, Jr. (December 31st, 1880 – October 16th, 1959) was an American statesman and soldier, famous for his leadership roles during World War II and the Cold War. He was Chief of Staff of the United States Army under presidents Franklin D. Roosevelt and Harry S. Truman, and served as Secretary of State and Secretary of Defense under Truman. He was hailed as the "organizer of victory" by Winston Churchill for his leadership of the Allied victory in World War II.

"The United States has recovered UFOs and their occupants. The UFOs were from a different planet and they were friendly. They have been hovering over defense facilities and airports. The U.S. authorities were convinced they had nothing to fear from them. The U.S. wanted people to concentrate on the real menace, communism, and not be distracted by the visitors from space. There has actually been contact with the men/aliens in the UFOs and there have been landings."

Vice Admiral Roscoe Hillenkoetter

Roscoe Henry Hillenkoetter (May 8th, 1897 – June 18th, 1982) was the third director of the post-World War II United States Central Intelligence Group (CIG), the third Director of Central Intelligence (DCI), and the first director of the Central Intelligence Agency created by the National Security Act of 1947. He served as DCI and director of the CIG and the CIA from May 1st, 1947 to October 7th, 1950 and after his retirement from the

United States Navy was a member of the board of governors of National Investigations Committee On Aerial Phenomena (NICAP) from 1957 to 1962.

NICAP

The National Investigations Committee On Aerial Phenomena was formed in 1956, with the organization's corporate charter being approved October 24th. Hillenkoetter was on NICAP's board of governors from about 1957 until 1962. Donald E. Keyhoe, NICAP director and Hillenkoetter's Naval Academy classmate, wrote that Hillenkoetter wanted public disclosure of UFO evidence.

Perhaps Hillenkoetter's best-known statement on the subject was in 1960 in a letter to Congress, as reported in The New York Times: "Behind the scenes, high-ranking Air Force officers are soberly concerned about UFOs. But through official secrecy and ridicule, many citizens are led to believe the unknown flying objects are nonsense."

"It is time for the truth to be brought out in open congressional hearings. Behind the scenes, high-ranking Air Force officers are soberly concerned about UFOs. But through official secrecy and ridicule, many citizens are led to believe the unknown flying objects are nonsense. To hide the facts, the Air Force has silenced its personnel through the issuance of a regulation."

Victor Marchetti

Victor Marchetti was the executive assistant to the Deputy Director of the CIA and is the co-author of "The CIA and the Cult of Intelligence," the only book ever censored by the US Government prior to publication up to 1979.

He lives in suburban Washington D.C.

My theory is that we have, indeed, been contacted– perhaps even visited– by extraterrestrial beings, and that the U.S. Government, in collusion with other national powers of the earth, is determined to keep this information from the general public. The purpose of the international conspiracy is to maintain a workable stability among the nations of the world and for them, in turn, to retain institutional control over their respective populations. Thus, for these governments to admit there are beings from outer space attempting to contact us, beings with mentalities and technological capabilities obviously far superior to ours, could, once fully perceived by the average person, erode the foundations of the earth's traditional power structure. Political, legal systems, religions, economic and social institutions could all soon become meaningless in the mind of the public. The national oligarchical establishments, even civilization as we know it, could collapse into anarchy.

Such extreme conclusions are not necessarily valid, but they probably accurately reflect the fears of the "ruling class" of the major nations, whose leaders (particularly those in the intelligence business) have always advocated excessive governmental secrecy as being necessary to preserve "national security."

The real reason for such secrecy is, of course, to keep the public uninformed, misinformed, and, therefore, malleable.

J. Edgar Hoover, head of the FBI

It appears J Edgar Hoover had admitted that the Army had recovered a downed UFO.

I would do it (study UFOs), but before agreeing to do it;

"We must insist upon full access to disks recovered. For instance, in the LA case the Army grabbed it and would not let us have it for cursory examination."

"The individual comes face-to-face with a conspiracy so monstrous he cannot believe it exists. The American mind has not come to a realization of the evil which has been introduced into our midst. It rejects even the assumption that human creatures could espouse a philosophy which must ultimately destroy all that is good and decent."

Harry Truman

President Harry Truman formed the highly secret group (code named MJ-12) which stood for Majestic-12. The sole purpose of this group was to investigate UFOs and report their findings to the president. Majestic-12 was established by a special executive order of President Truman on September 24th, 1947, three months after the crash at Roswell, New Mexico.

A briefing document describing the group for incoming president Dwight Eisenhower was leaked and is now available to almost anyone. The government denies the existence of such a group or document. The authenticity of the document cannot be proven, but the reality and the purpose of MJ-12 has been confirmed with many high-level sources over the years.

"I can assure you that flying saucers, given that they exist, are not constructed by any power on earth."

President John F. Kennedy in 1963

"I'd like to tell the public about the alien situation, but my hands are tied."

President John F. Kennedy to Bill Holden, steward on Air Force One, flying over Germany, summer 1963 When Holden asked Kennedy what he thought about UFOs.

Chapter 10

Insight

The meaning of insight in medical terms is powerful.

In psychiatric terminology one who has lost insight to his or her predicament is actually has a major psychiatric disorder.

It is the marker of insanity.

Most of the other disorders with insight into the plight of an individual in a given stressful situation are minor psychiatric disorders.

In depressive illnesses, one who has lost insight to the given predicament is said to have Endogenous Depression.

One who has insight into the predicament is called Reactive Depression.

Insight into a scientific problem is what propels the science forward.

What it means is given a certain scientific problem, one has to delve into its core and bring out a meaningful explanation for the occurrence of a particular phenomenon.

Often there are no measurements or data.

A hypothesis is developed to explain the phenomenon and thereafter it is tested for its validity.

Insight by its nature predates the hypothesis.

Insight can be of used outside the scientific arena and may herald clever resolution of complicated conflicts, especially political. We could say Nelson Mandela's approach to apartheid showed his mature political insight.

Mental Culture

Mind is the most powerful weapon (that is how Americans use Mind Culture) in this world.

It can be used for good and bad.

In practice it is used as a weapon and psychological warfare in politics and religion.

Secret services (CIA, MI5, FBI, KGB) are examples of the clandestine use of this psychological warfare in the past century.

Only in Buddhism or Hindu Culture, it is used for benevolence. Christianity (fear of god) and Muslim (coercion to inculcate fear and conversion) religions use this fear psychosis for expansion of their control.

One need not fear but develop the mental culture of Metta from an early age.

But what is promoted is Hate Culture.

The tolerance vanishes with this psychosis.

One should have the freedom not to be indoctrinated by any pseudo-religion.

Metta Meditation is something that everybody (it takes the fear psychosis out of your mind) can practice.

I prefer the Moment Meditation.

There is a caveat here.

One should not practice these mental techniques in old age without a suitable meditation master.

One might end up psychologically deranged or broken down. I have seen several in my life.

Ajan Chach (in the past) and Ajan Brakmavaso (current) are good exponents (guides).

Ajan Brakmavanso to his credit has many simple books in English. His talent is that he can put very difficult context in very simple words using a simile. There is nobody to match him currently.

Meditation culture needs hard and strenuous practice.

One cannot master this in a weekend course.

If somebody is doing academic work "Vissusdhi Magga" is a colossal text. That is the only book I do not possess. One need only one or most two to three techniques depending on the personality.

For me "Dhamma-Pada", the small volume with 400 odd verses is adequate as an introduction. Even from that I do use not more than 10 verses for insight.

The Pali word for "Insight" is "Vipassana", which is being adopted by Buddhist practitioners of meditation.

I prefer the term "Mindfulness Meditation".

I go, little further to "Moment Meditation", which is Mindful presence every wakeful moment of living (except one is in dream state of sleeping).

There are different mental cultures from "Mediums" to "Channels" to "Insight Meditation".

Last is the most difficult.

There is a lot to pick but one needs a proper Meditation Master to guide one through.

Chapter 11

Imagination

Children are born with imagination to make up for their lack of conceptualization of a given situation. Once one becomes an adult and able to conceptualize a given problem (at his or her level of competence), unfortunately, one loses this ability to think imaginatively.

The man's creativity lies in this faculty of imagination.

The religion uses this transition period (lack of creative thinking ability) to insinuate bizarre concepts.

The god created this world is one such imaginary concepts. I do not blame anyone for looking up, at the sky, for an imaginary god and then find stars as gods.

The concept of god is an example of human capacity for imagination. There is nothing wrong in it but this imaginary world is put as a dogma, the problems start to crop up. More and more dogmas are put in this realm, often without insight.

The scientific reasoning is muted from the very beginning.

It becomes a dogma when enough followers take up the same conviction. It leads to a belief system and there is no end to its influence and expansion, very often not tested for credibility.

Unfortunately, "Big Bang Theory" of creation of the universe has fallen into this trap. We have not put an alternative theory to counteract it. This has made the science to ignore the Dark Matter and its hidden physics.

In material world one can use imagination to create an end product and one can test it for its suitability (fit for purpose).

But in the immaterial world or paranormal psychology there is no way of testing the empirical (*verifiable by observation or experience rather than theory or pure logic*) experience.

It is an individual experience, not verifiable by a third party.

One avenue to verify, if something is imagined or real is insight into the given predicament. It has to be done by the individual who has imagined it, in the first place (with average intelligence).

What the individuals with higher intelligence should do is to help them into insightful thinking. What has happened in history is the exact the opposite.

We have more people with average intelligence and only a minority with superior intelligence. The intelligent lot knows the predicament of the less intellectual soul and coerced them into either realm of fear or dogma.

The end point of imagination is the creation of a dogma.

That is something worthwhile avoiding.

Keeping an open mind is the exit strategy.

Chapter 12

Conversion

Lack of imagination and insight leads to a state of mind called conversion. It is a fix or rigid state of mind filled with prejudices, be that it may, religion, nationality, race, caste, creed or any other cultural dogma.

I believe this is where the cultural instinct get ingrained.

If a particular population is isolated and not exposed to the rest of the world civilizations, and when there are many such isolations, it is easy for many dogmas to spin off independent of the others.

Man's irresistible urge for migration has brought these dogmas face to face with conflicts. As a result of these conflicts some dogmas disappear and some get established.

This is probably residing in our genome.

Religion is one exit point.

The black and white race is another universal dogma.

Race is another.

All are end points of conversion strategy of mankind.

Any other dogma falls within this frame work.

The existence of god and a creation of an almighty wisdom are byproducts.

Once inside this dogma, science takes a back seat.

Albert Einstein put this succinctly.

Science without religion is lame and religion without science is blind.

Chapter 13

Basic tenet of Science

Apart from methodology, the framework of science is based on validation of facts from fiction.

To begin with it is an insight into a problem and formulating a hypothesis.

Inquiry and validation come later not necessarily by the author of the hypothesis but often by a different set of individuals.

Let me propose a hypothesis "That we are not alone in this universe".

That sets in the formation of a forum of journalists, scientists and citizenry to investigate the problem at hand. The idea is not to speculate or proclaim with religious zeal but to scientifically prove or disprove the stated claims.

What has happened in history was to cover up the facts and lead astray the public and preparing the breeding ground for conspiracy theory.

The result is various people with different academic credentials take up the challenge to validate their own positioning and reasoning.

From scientists to engineers to journalist, get involved in unraveling the mystery.

For the citizenry it is a mystery.

It's highly unlikely we're alone in the universe, NASA experts are saying and we may be close to finding alien life.

In fact, it may happen in the next two decades.

NASA held a panel discussion at the agency's Washington headquarters where space experts talked about the search for Earth-like planets that host life.

Based on recent advancements in space telescope technology, scientists estimated that in the coming decades we'll confirm suspicions that we are not alone.

"I think in the next 20 years we will find out we are not alone in the universe," NASA astronomer Kevin Hand believes.

NASA Administrator Charles Bolden echoed Hand's sentiment.

"It's highly improbable in the limitless vastness of the universe that we humans stand alone," he said.

Recently, NASA's Kepler Space Telescope picked up on an Earth-like planet in the "habitable zone" of another star. At the time, the observation of the planet, Kepler-186f, was hailed as the first discovery of an Earth-size planet orbiting in the habitable zone of a star similar to our sun.

Scientists believe there are potentially many more Earth-like planets in the universe — and some of them could be home to alien life.

"Astronomers think it is very likely that every single star in our Milky Way galaxy has at least one planet which can support life", Sara Seager, professor of planetary science and physics at Massachusetts Institute of Technology, said during the talk.

"Sometime in the near future, people will be able to point to a star and say, 'that star has a planet like Earth.'

With the expected launch of the James Webb Space Telescope in 2018, NASA's planet-hunting mission will get an extra boost. The new piece of equipment is designed to study infrared light, making it easier to spot extra-solar planets.

But NASA may need even larger and more powerful telescopes to discover alien life.

"To find evidence of actual life is going to take another generation of telescopes," Matt Mountain, director of the Space Telescope Science Institute, said at the event.

"And to do that, we're going to need new rockets, new approaches to getting into space, new approaches to large telescopes — highly advanced optical systems."

What I find intriguing is that American Scientists (including perhaps NASA) took over 70 years to reveal the facts they had within their domain. Waiting so long, until the development of the correct telescope with enough resolution was not the only option.

I am referring to Roswell incident.

If they reveled it 70 years ago, the world would have been entirely different altogether, today.

The flimsy argument put forward does not hold water.

I contribute this to either lack of insight and foresight, outright stupidity or grand conspiracy to create a hidden agenda of science (what they already knew first hand from aliens).

Chapter 14

How We Killed Our Science Education

It was a strange coincidence that I took to science for higher education.

There were no mentors.

There were no specialists to introduce it in smaller doses. If not for the British Council Library, I would have gone astray. There were plenty of booklets from photosynthesis, to life on earth, to other scientific trends, at that time. Everything was changing rapidly in nineteen seventies, from scientific fiction films to science.

The school library did not have anything of substance. Luckily school times were short and after school, I went to the British Council library and glanced through the science magazines.

They were not for burrowing.

Few years before, the government at the behest of stupid politicians decided to change the medium of instruction. Luckily, I had the option to select English as the medium of instruction. My English was no good but I knew given the time, I will pick up the terminology in science.

There were no dictionaries in science. The science teachers without training in teaching would dish out the notes they collected in their education ladder. To be precise, the university notes which they never understood.

We were the guinea pigs.

But books in the British Council Library on science and colourful magazines kept me ahead of time.

When the medium was changed, it took more than ten to fifteen years for the books to be translated from English. It took few more years for them to be published. When the books were finally dished out to the students they were hopelessly outdated.

This is how we killed the science education in schools.

There were attempts to teach Medicine in mother tongue.

I was a raw hand taken in by the University, mainly to translate and teach the translated stuff verbatim.

I craftily made my professors to understand the gravity of it and subtle change in meaning after translation. I told them it was not worth the effort. Even after three months of deliberations, if one of the members found a suitable word, I would tear it off to pieces with counter arguments.

Finally we decided to call it off.

Higher education was in English but there was option for one to select mother tongue if one wished.

It worked in seventies.

I do not know what is the trend now.

Chapter 15

Why Science took a back seat in Human History

The biggest obstacle to science was the religion.

That includes Buddhism, too.

Even though Buddhism encouraged Scientific Inquiry in "Kalama Sutta", it was never taken forward in the Eastern culture.

When religion takes its roots dogmas get embedded in human thinking.

But when we consider how religion originated, the answer becomes startlingly curious.

Man had an affliction to mood elevating drugs from prehistoric time. Mood elevating drugs impair critical thinking capacity. Some of these psychedelic drugs put man in "Trance States."

Unfortunately, some started believing these trance states as real messages from god or heaven.

Some of these guys in the human hierarchy who were addicted to these potent substances for their existence and survival created bizarre theories of human origin, including God. On top of the religion there is an another dogma, God and his creation.

In actual fact, it is these psychedelic drugs that retarded critical thinking and the evolution of scientific thinking.

The bottom line is the drugs that effect sleep, memory and critical thinking are the byproducts of chemistry that came with scientific evolution.

In actual fact, science itself started killing its roots.

LSD is a case in point.

Alcohol too effects memory and normal sleep.

Imagine an airline pilot or an astronaut who is addicted to mood elevating drugs.

Would you board on a spaceship manned by an addict?

For our critical capacity to work efficiently, clear mind and perfect memory are vital ingredients.

Our survival in this century depends on these factors.

Any political leader with paranoid inclination would derail human survival.

I would summarized my thinking below.

Right Constitution

1. Scientific outlook comes first.

2. Rules of Law and fair play and Justice to all living beings (Plants and Animals included) is second.

3. Democracy as a force and the tolerance of diverse views is the third.

4. Universal suffrage without bondage, dogma and external powers influencing (encourage the internal powers of the mind) the decision making, is the fourth.

Religion has no place or should reign below the above four pillars. If an individual has a certain conviction for solace, be that be so, but he/she should tolerate views of the others including, especially those who want not to be bound by any religious dogma.

Chapter 16

Conspiracies in Science

I thought I won't be able write a book on "Conspiracies in Science."

I thought I could only make three to five conspiracies with some hard soul searching.

But it does look like present day scientists are hell bent on creating conspiracies. It looks like they thrive on conspiracy.

Base Area 51 in USA is a case in point.

I accidentally stumbled upon it.

It took considerable time and effort to unravel, the scenario left behind by Roswell accident.

I am not sure which is a fact and which is a fiction, as far as Roswell saga is concerned.

It is shrouded in mystery.

This is, in spite of critical analysis.

It is very difficult to break down a complex scientific problem in simple steps. The way a scientist would try to solve a given problem is methodical and rational. It could be very simple or very difficult. A scientist would try to break the known facts to simpler components.

If scientists can be incriminated in cover up and misinformation, it becomes much harder to unravel the truth as well as credibility.

When scientists, secret service and politicians become a cohort in covert operation, even though the initial intent may be very benign to begin with, but with time, it leads to speculation, sinister design and conspiracy.

I lost count of many conspiracies in the last century, and of course, I left out the minor events. To begin with I must say this is not a comprehensive analysis but a limited attempt. The idea is to stimulate the reader so that he/she can do one's own study.

I prefer to give the benefit of doubt to the scientists, even in the case of Area 51.

Let's see that overwhelming concern was local and international security, assuming it was a nuclear testing facility.

Then the ramifications are many.

Let us say the scientists did not know all the health risks.

The Radium and Mary Curie is a case in point.

Let's for argument sake validate that the scientists knew all the implications and kept all the civilians out of reach of harm.

Let's postulate that this was a grand game plan to activate the private sector companies to take the lead in research (reverse engineering) and the government to become only a facilitator and a protector of civilian interest. The private sector companies needed top security from competitors from the outside world, at the time of the cold war.

Cold war itself could be a cover up for build up of various arsenal.

Let's consider the test aircrafts were flying with or without nuclear warheads and any mishap would be an international disaster.

Let's extend it as a welcome condition.

Let's assume the alien visitors with higher technology did arrive and made contact.

Let us surmise, the sole purpose of them visiting the earth is to find a solution to their disease vulnerability.

The final conclusion, with their superior intellect, they formulated that they needed a guinea pig (human beings in this case) to test.

They will barter technology for such an exchange of material both physical and biological.

Just look at the words I have used, say, see, postulate, consider, assume and surmise.

None of those word will stand in a scientific discussion.

They are all in a literal sense, an opinion or an expression.

The end result is speculation and misrepresentation.

The list can go on but with it, the public curiosity will build up, for free information, for which no liberal government could have, neither the control nor the defense of its secrecy.

So the project managers have to devise plausible contingencies.

Are there any other possibilities?

Are there any other probabilities?

I started with scientific intent but ended up with probabilities but not possibilities.

That was not what I intended.

So the word conspiracy has to stay put.

Microsoft's Strategy

Let us begin with Microsoft, a relatively harmless business venture and how the Microsoft developers set about its expansion.

Their strategy is simply monopoly.

They would use the FUD, the Fear, Uncertainty, Doubt in the minds of gullible users.

Bulldoze other operating systems with Fear, Uncertainty and Doubt (FUD).

Then cover up all the loose ends and vulnerabilities.

Allow the use of pirated copies of a basic system with total vulnerability to viruses. Then once it is in free circulation threaten to take legal action to sue the users having pirated copies.

The user is addicted to Windows by now with games as a lure and he/she has no option but to buy the revised version.

I was very much into Windows and I made sure I always had a paid up copy. Then without my knowledge a virus ate into the system. Within 24 hours, I detected it but by that time it was too late it had infected my sons computer which he used for games. Then I went for the virus treatment and it slowed the system and playing games became tedious.

I accidentally, stumbled upon copies of Demo Linux Knoppix and Damn Small Linux.

I was hesitant but having read the book "The Joy of Linux", written in the style of "Joy of Sex", by Peter Parfitt and Jon (Maddog) Hall, I plunged into the deep secrets of Linux.

In no time, I made vow that "I will never use Windows again".

I was convinced Microsoft had a big conspiracy, in their shadow in addition to their monopoly.

Beside, why pay money for a poor operating system.

However, Microsoft had a deal with game developers and that was the real reason for their success, not computing but entertainment.

While Linux was far ahead of computing, resource management, and the cloud computing, they did not have enough games to lure the user. I was very open about this deficiency all along but now there is a distribution dedicated for games called Steam Linux.

It is an open secret Microsoft uses Linux at its back end especially in the "Cloud".

Not everybody is hell bent on conspiracy.

There are good guys too.

I will deal with them later.

Chapter 17

Drug Companies

Chocolates and Drugs

Running parallel with the demise of Prof. Senaka Bibile and Mr. Upali Wijewardane, the business magnate, is not appropriate but I was enjoying some Sri-Lankan chocolate, it struck a code of concordance, in brief moment of realization that both of them took over big companies whose Head Quarters (HQ) are grafted in Switzerland under cover of proprietary rights.

It is now known chocolates are good, godsend source of pharmaceutical ingredients that improve cognitive function of the young and the old.

That would have been the reason for me hitting this brilliant idea of running parallels.

Let me list a few of them and the reader can do his or her own research.

1. The original chocolate recipe is still kept under top secret dossier and nobody can copy it except one big company registered there.

2. All big drug companies have there HQ in Switzerland and covered by Copyright Law.

These companies have unrestricted opportunity to make money not in millions but in billions and deposit part of that money in Swiss Banks and under the cover of secrecy and are not required to pay taxes.

3. This is how the capitalist instruments work to protect big guns including Gun Lobbyists.

4. Coming to Mr. Upali Wijrardena, he was the only guy who had the guts and the brains to produce a chocolate not originating from Switzerland.

5. Everybody knows how Prof. Senaka Bibile under the aegis of UNDP wanted to promote non-proprietary drugs.

6. Both demised in midair, in mid seventies under mysterious circumstances.

7. I believe the plot was hatched to get rid of them in America and the planning was done in Switzerland.

8. These are the very guys under UNO asking transparency from emerging small nations.

9. I think the Transparency International should open the dossiers on these two deaths and investigate the true facts for the sake of the history of the Capitalistic World.

10. In the mean time any Sri-Lankan who is short of ideas and wants to investigate forensic evidence how these events unfolded in nineteen seventies should do so with his or her own thesis proposal, bearing in mind nobody would come forward to finance the research.

11. If anyone does it for the sake of history, he/she makes a good chapter in history and might even get a generous grant from the very instruments who were instrumental in trying to reverse the emerging trends of yesteryear. However, with a minor

proviso that some of the informations have to be suppressed by default.

While these were happening our guys have produced, an array of chocolates with a better taste than those originating from India and Switzerland with no help from the big guns.

In the mean time, all chocolate lovers should eat only the Sri-Lankan brands but not imported brands which are based on Swiss Brands but made in India with very poor quality ingredients.

Instead of writing about particular conspiracy of a particular company, brief review of the life of Late Professor Senaka Bibile is in line.

Late Professor Senaka Bibile

The notes below is a reproduction.

The writer is also a teacher of mine.

He was teacher and a role model of yesteryear.

Reproduced here is a short review, for present doctors to read who are pampered by drug companies with money spinning around.

The Peradeniya Medical Faculty celebrated 50 years in January 2012. This article reflects some aspects of the Medical Faculty and its work.

The purpose of this short piece is to document the beginnings of the study of medical education in Sri Lanka and to acknowledge Senaka Bibile's role in initiating it.

Senaka Bibile (1920-1977) the person and his work, are well known in medical, political and Intellectual circles within Sri Lanka. I wanted an international perspective on this remarkable man. Hence I scoured the Internet and came across this anonymous quote in Wikipedia;

"He was the greatest medical benefactor of humanity that Sri Lanka has hitherto produced."

The thoughts and the subsequent work which earned him this reputation occurred during his time at the Peradeniya Medical School (PMS) as Professor of Pharmacology (1962-1967) and subsequently, as its first Dean (1967-1970).

I knew Senaka Bibile fairly closely, as a friend, as his occasional physician and as a collaborator in the field of medical education. Senaka's work in this field is not as well known as his work in helping nations set up national drug policies and in promoting the rational use of drugs. When Bibile decided to move from congested Colombo to the free thinking environment of the Peradeniya Campus, he was brimming with ideas for making the PMS as a medical school based on rational approaches to education and on scientific thinking. With this in view, he wanted himself and the teaching staff to learn about the process of training doctors. Using his charisma and affable personality he invited the staff to his own house in the Campus on Tuesday evenings (after the 'Tuesday Talks' which was a lecture series for staff and students, and a precursor to the Kandy Society of Medicine) to read and discuss a book on medical education titled "Teaching and Leaning in Medical School" by Stephen

Abrahamson and George Miller. Many staff readily accepted his invitation. Each one of us took on the responsibility for reading and verbally summarizing a chapter from this book. These informal meetings were held in a spirit of camaraderie and were thoroughly enjoyable. (Later in 1973, under the auspices of the Medical Education Unit, the authors of this book conducted a Teacher Training Workshop for teachers from the PMS and the Colombo Medical School). After several of these Tuesday evening sessions a formal 'Working Group on Medical Education" (WGME) was established in the PMS.

The 1960s was also the time when the World Health Organization was getting interested in Medical Education (It used the wider term of Health Manpower Development) and was willing to support any local efforts made to train teachers in medical schools in educational science. Noting the efforts which the PMS was making in this field it started to provide support for medical teachers to strengthen their knowledge and skills in medical education which finally resulted in the establishment of the Medical Education Unit in 1973, the first such Unit in the country and in the South East Asian Region.

Senaka Bibile's implantation of the idea of medical education in the PMS has now grown and reproduced itself to become a well established discipline in medicine, on par with any other discipline in the medical field. All medical schools and many schools which train health personnel in the island now have a Medical Education Unit or its equivalent.

by Professor T. Varagunam

Cholesterol and its treatment

I had been teaching atherosclerosis all my life and I never believed that one need to lower blood cholesterol with drugs.

Cholesterol, from the Ancient Greek chole- (bile) and stereos (solid) followed by the chemical suffix -ol for an alcohol.

It is a vital organic molecule of life.

It is a sterol (or modified steroid), a lipid molecule and is biosynthesized by all animal cells. It is an essential structural component of all animal cell membranes that is required to maintain both membrane structural integrity and fluidity. Cholesterol enables animal cells to dispense with a cell wall (it protects membrane integrity and cell viability) thus allowing animal cells to change shape and animals to move (unlike bacteria and plant cells which are restricted by their cell walls).

In addition to its importance for animal cell structure, cholesterol also serves as a precursor for the bio-synthesis of steroid hormones and bile acids.

Cholesterol is the principal sterol synthesized by all animals. In vertebrates the hepatic cells typically produce greater amounts than other cells. It is absent among prokaryotes (bacteria and archaea), although there are some exceptions such as Mycoplasma, which require cholesterol for growth.

François Poulletier de la Salle first identified cholesterol in solid form in gallstones in 1769. However, it was not until 1815 that chemist Michel Eugène Chevreul named the compound "cholesterine".

Cholesterol is oxidized by the liver into a variety of bile acids. These, in turn, are conjugated with glycine, taurine, glucuronic acid, or sulfate. A mixture of conjugated and non-conjugated bile acids, along with cholesterol itself, is excreted from the liver into the bile. Approximately 95% of the bile acids are reabsorbed from the intestines, and the remainder are lost in the faeces.

The excretion and reabsorption of bile acids form the basis of the enterohepatic circulation, which is essential for the digestion and absorption of dietary fats.

Under certain circumstances, when more concentrated, as in the gallbladder, cholesterol crystallises and is the major constituent of most gallstones.

For me cholesterol is a central molecule in biochemistry which produces many compounds including the formation of sex hormones. It is universally present in all cells. In fact, cells and cell membranes including the nerve cells cannot function without it. The liver has a mechanism to preserve cholesterol and excrete it only when oxidized or metabolically deranged.

One must note that macrophages (big fat cells that scavenge unwanted material) do have various receptors to ingest both normal and oxidized cholesterol. Its receptor for oxidized cholesterol is called literally a non-specific scavenger receptor and once taken in unable to recirculate and fat with its cholesterol begin to build up within the cell. Whereas the normal cholesterol bound to universal receptor can be made available and recirculates.

Basic problem in cholesterol metabolism is oxidation injury and the treatment should be antioxidants not anti-cholesterol drugs.

It is the scientific rationale.

Entero-hepatic circulation and bile acids secretion by the liver make sure cholesterol is conserved.

Bile acids of course help in digestion of fats.

In my opinion cholesterol cannot be used as a health / disease indicator for many reasons.

Its levels are determined, not only by diet but also by basic biochemical pathways in intermediary metabolism.

In my opinion any lovering of it will have untoward effects.

Now take the drug Simvastatin and its derivatives which effect the hepatic dehydrogenase.

It does cause liver damage by inhibiting normal biochemical pathways.

It is toxic to the liver.

After 30 years of its use now they are reporting the relationship of Senile dementia to low level of cholesterol. Senile dementia is a debilitating disease and the sufferers cannot take libel action against the company which marketed it. It also causes muscle weakness which is a major handicap in old age.

I have used only one example but there are thousands of pharmaceuticals that cause side effects as the direct result of their use. The bottom line is drug conspiracy. My advice is, "one should avoid all drugs with side effects".

Chapter 18

Zoonosis

Zoonoses (also spelled zoönoses; singular zoonosis (or zoönosis); from Greek word "animal" and "sickness") are infectious diseases of animals (usually vertebrates) that can naturally be transmitted to humans.

Major modern diseases such as Ebola virus disease, salmonellosis and influenza are zoonoses. Zoonoses can be caused by a range of disease pathogens such as viruses, bacteria, fungi and parasites. Of over 1,500 pathogens known to infect humans, nearly half are zoonotic.

Zoonoses have different modes of transmission.

In direct zoonosis the disease is directly transmitted from animals to humans through media such as air (influenza) or through bites and saliva (rabies).

In contrast, transmission can also occur via an intermediate species (referred to as a vector), which carry the disease pathogen without getting infected.

When humans infect animals, it is called reverse zoonosis or anthroponosis.

This preamble is necessary before opening up a debate on conspiracy theory. In the late 19th century and the beginning of the 20th century, there was an awakening in the Western world both in scientific terms and politically sense. While this was happening in the West, the primitive societies were in abject poverty and were under colonial Western influence. There were no human rights charters but human right violation was the vogue in the name of an attempt to civilize these primitive people.

West and Mid Africa especially Congo is a very good example under Belgium King Leopoldville.

When a particular civilization get isolated in an ecological niche, there are certain characteristics built into the system. These characteristics change very little with the population expansion and migration (from an unfertile niche to an adjoining fertile niche, not occupied by an alien civilization).

But when a complete foreign or alien civilization encroaches upon these primitive civilizations catastrophe ensues.

First reaction is overt and covert war.

Then it leads to either domination by the aliens or succumb to the primitives.

Numbers invading are small by default but, if they have advanced technology (guns to begin with) there is no defense for the primitives.

This is what happened to Native Americans with Spanish invasions and later African continent by German, French, British and Belgians.

These invaders bring diseases to the natives which varied from syphilis, to all venereal diseases, to measles, to aids.

The sars epidemic is different.

Usually there is no man to man transfer in zoonosis. It is usually animal to man and it terminates either with the demise or of the recovery of the human individual.

When some of these viruses mutate and acquire variants that can transfer infection from man to man, pandemic results. The zoonotic variants in the modern world usually originate from fowl or pigs or domesticated animals, in very poor living conditions.

Pandemics can wipe out entire civilizations.

Pandemics were used by alien migrants to decimate the natives.

The next wave in this design is the insinuation of volunteer medical personal (with putative medications), as healers.

When this strategy fails the divine visitors or priests with abundance of money often amassed by ill gotten means replace "the medicine sans frontiers".

Colonial Expansion

Leopold II (9th April 1835 – 17th December 1909) was the second King of the Belgians, known for the founding and exploitation of the Congo Free State as a private venture. Born in Brussels as the second (the eldest surviving) son of Leopold I and Louise of Orléans, he succeeded his father to the throne on 17th December 1865, reigning for exactly 44 years until his death. This was the longest reign of any Belgian monarch.

Leopold was the founder and sole owner of the Congo Free State and all its people, a private project undertaken on his own behalf. He used explorer Henry Morton Stanley to help him lay claim to the Congo, an area now known as the Democratic

Republic of the Congo. At the Berlin Conference of 1884–1885, the colonial nations of Europe authorized his claim by committing the Congo Free State to "improving the lives of the native inhabitants".

From the beginning, however, Leopold ignored these conditions and millions of Congolese inhabitants, including children, were mutilated and killed. He ran the Congo using the mercenary Force Publique for his personal enrichment. He used great sums of the money from this exploitation for public and private construction projects in Belgium during this period. He had to donate these private buildings to the state before his death.

Leopold extracted a fortune from the Congo, initially by the collection of ivory, and after the rise in the price of rubber in the 1890s, by forced labour from the natives to harvest and process rubber. Under his regime millions of Congolese people died; modern estimates range from one to fifteen million, with the consensus figure around ten million.

Human rights abuses under his regime contributed significantly to these deaths. Reports of deaths and abuse led to a major international scandal in the early 20th century, and Leopold was ultimately forced by the Belgian government to relinquish control of the colony to the civil administration in 1908.

Exploitation and atrocities

Leopold amassed a huge personal fortune by exploiting the natural resources of the Congo. At first, ivory was exported, but this did not yield the expected levels of revenue. When the global demand for rubber exploded, attention shifted to the labor-intensive collection of sap from rubber plants. Abandoning the promises of the Berlin Conference in the late 1890s, the Free State government restricted foreign access and extorted forced labor from the natives. Abuses, especially in the rubber industry, included the forced labor of the native population, beatings, widespread killing, and frequent mutilation when the production quotas were not met. Missionary John Harris of Baringa, for example, was so shocked by what he had come across, that he wrote to Leopold's chief agent in the Congo saying:

"I have just returned from a journey inland to the village of Insongo Mboyo. The abject misery and utter abandon is positively indescribable. I was so moved, Your Excellency, by the people's stories that I took the liberty of promising them that in future you will only kill them for crimes they commit."

Estimates of the death toll range from one million to fifteen million, since accurate records were not kept. Historians Louis and Stengers in 1968 stated that population figures at the start of Leopold's control are only "wild guesses", and that attempts by E. D. Morel and others to determine a figure for the loss of population were "but figments of the imagination".

Mutilated people from the Congo Free State

Adam Hochschild devotes a chapter of his book King Leopold's Ghost to the problem of estimating the death toll. He cites several recent lines of investigation, by anthropologist Jan Vansina and others, that examine local sources (police records, religious records, oral traditions, genealogies, personal diaries, and "many others"), which generally agree with the assessment of the 1919 Belgian government commission: roughly half the population perished during the Free State period. Hochshild points out that since the first official census by the Belgian authorities in 1924 put the population at about 10 million, these various approaches suggest a rough estimate of a total of 10 million dead.

Smallpox epidemics and sleeping sickness also devastated the disrupted population. By 1896, African trypanosomiasis had killed up to 5000 Africans in the village of Lukolela on the Congo River. The mortality statistics were collected through the efforts of British consul Roger Casement, who found, for example, only 600 survivors of the disease in Lukolela in 1903.

Medicine, Science and Germans in Mid West Africa

In her African and South Sea colonies Germany established diverse biological and agricultural stations. Staff specialists and the occasional visiting university group conducted soil analyses, developed plant hybrids, experimented with fertilizers, studied vegetable pests and ran courses in agronomy for settlers and natives and performed a host of other tasks.

Successful German plantation operators realized the benefits of systematic scientific inquiry and instituted and maintained their own stations with their own personnel, who further engaged in exploration and documentation of the native fauna and flora.

Research by bacteriologists Robert Koch and Paul Ehrlich and other scientists was funded by the imperial treasury and was freely shared with other nations. More than three million Africans were vaccinated against smallpox. Medical doctors the world over benefited from pioneering work into tropical diseases and German pharmaceutical discoveries "became a standard therapy for sleeping sickness and relapsing fever.

The German presence (in Africa) was vital for significant achievements in medicine and agriculture.

In other words these primitive people were the subjects or in real terms the human guinea pigs.

AIDS Epidemic

HIV-1 from chimpanzees and gorillas to humans

Scientists generally accept that the known strains (or groups) of HIV-1 are most closely related to the simian immunodeficiency viruses (SIV) endemic in wild ape populations of West Central African forests. Particularly, each of the known HIV-1 strains is either closely related to the SIV that infects the chimpanzee subspecies or closely related to the SIV that infects western lowland gorillas.

The pandemic HIV-1 strain (group M) and a very rare strain only found in a few Cameroonian people (group N) are clearly derived from chimpanzee populations living in Cameroon. Another very rare HIV-1 strain (group P) appear to derive from gorilla populations of Cameroon. The primate ancestor of HIV-1 strain (group O) is a strain infecting over 100,000 people mostly in Cameroon but also the neighboring countries.This has been recently confirmed to be derived from gorilla strain.

The pandemic HIV-1 group M is prevalent in the southeastern rain forests of Cameroon (modern East Province) near the Sangha River.

Thus, this region presumably is where the virus was first transmitted from chimpanzees to humans. However, reviews of the epidemiological evidence of early HIV-1 infection in stored blood samples, and of old cases of AIDS in Central Africa have led many scientists to believe that HIV-1 group M early human center was probably not in Cameroon, but rather farther south in the Democratic Republic of the Congo, more probably in its capital city, Kinshasa (formerly Léopoldville).

Using HIV-1 sequences preserved in human biological samples along with estimates of viral mutation rates, scientists calculate that the jump from chimpanzee to human probably happened during the late 19th or early 20th century, the time of Western exploitation, rapid urbanization and colonization in equatorial Africa.

Exactly when the zoonosis occurred is not known.

Some molecular dating studies suggest that HIV-1 group M (most recent common ancestor) started to spread in the human population, in the early 20th century, probably between 1915 and 1941.

This happened in between the two World Wars, I and II when Germans used the African bases, especially the Democratic Republic of Congo as an experimental base for clandestine medical research. They probably used cells derived from simian kidney for cell culture studies. What was studied by Germans scientists are not known after their fall but it looks like American hijacked the project to America with captured German scientists.

What is now known for certain and the researchers at that time failed to recognize was the accidental infection of simian virus in human (African volunteers were used as guinea pigs) subjects.

There were no established protocols or safety concerns for decades. The transfer of infection most probably occurred through infected unsterilized needles and syringes.

The aids virus is somewhat similar to hepatitis B virus and it is quite possible that it could carry the virus from man to man (in living lymphocytes) unless something happens in the intestinal system of the mosquitoes. It appears that the mosquito saliva and intestinal system inactivate both the virus and the infected lymphocytes.

Additionally, the variants of theses viruses originated in cell culture media, I believe. I guess the scientist were trying to find a less infective variant similar to Small Pox (animal variant used for vaccination) variant that can be used for vaccination.

This experiment went haywire.

Direct inoculation by unsterilized injection needles was the primary mode of infection and not zoonosis. Ignorance of the natives and their abject poverty made it easy for the colonial masters to hide the facts and figures.

In a way it was a grand conspiracy.

I feel that the biological scientists of the 21st century should study the plausible cover ups.

What intrigued me at the beginning of the aids epidemic was why the adult mosquitoes did not spread the disease?

It (mosquito) did not but the syringes did.

A study published in 2008, analyzing viral sequences recovered from a recently discovered biopsy made in Kinshasa, in 1960, along with previously known sequences, suggested a common ancestor transmitted to humans between 1873 and 1933 (with central estimates varying between 1902 and 1921).

The primates being a critically endangered species, further confirmatory research is impractical. However, the researchers were able to hypothesize a phylogeny from the limited available data. They were able to use the molecular clock of a specific strain of HIV to determine the initial date of transmission, which is estimated to be around 1915-1931.

HIV-2 from sooty mangabeys to humans

Similar research has been undertaken with SIV strains collected from several wild sooty mangabey populations of the West African nations of Sierra Leone, Liberia, and Ivory Coast. The resulting phylogenetic analyses show that the viruses most closely related to the two strains of HIV-2 had crossed the species barrier to spread in humans subjects. The ancestral virus was (current variants are HIV-2 groups A and B) found in the sooty mangabeys of the Tai forest, in western Ivory Coast.

There are six additional known HIV-2 groups with limited or dead end penetration.

Molecular dating studies suggest that both the epidemic groups (A and B) started to spread among humans between 1905 and 1961 (with the central estimates varying between 1932 and 1945).

Bushmeat Hypothesis

According to the natural transfer theory (also called 'Hunter Theory' or 'Bushmeat Theory'), the "simplest and most plausible explanation for the cross species transmission" of SIV or HIV (post mutation), the virus was transmitted from an ape or monkey to a human when a hunter or bushmeat vendor/handler was bitten or cut while hunting or butchering the animal. The resulting exposure to blood or other bodily fluids of the animal can result in SIV infection.

Prior to World War II, some Sub-Saharan Africans were forced out of the rural areas because of the European demand for resources. Since rural Africans were not keen to pursue agricultural practices in the jungle, they turned to non domesticated meat as their primary source of protein. This over exposure to bushmeat and unhygienic practice of butchery increased blood-to-blood contact and increased of transmission.

A recent serological survey showed that human infections by SIV are not rare in Central Africa.

How the SIV virus would have transformed into HIV after infection of the hunter or bushmeat handler from the ape/monkey is still a matter of debate, although natural selection would favor the establishment on the new host.

The virus reproduces in the T cells.

A study published in 2009 also discussed that bushmeat in other parts of the world, such as Argentina, may be a possible location for where the disease originated. HIV-1C, a subtype of HIV, was theorized to have its origins circulating among South America. The consumption of bushmeat is also the most probable cause for the emergence of HIV-1C in South America.

However, the types of apes/monkeys, shown to carry the SIV virus were not determined. The monkey species are different in South America. The primary point of entry, according to researchers, is somewhere in the jungles of Argentina or Brazil. An SIV strain, closely related to HIV, was interspersed within a certain class of primates. This suggests that the zoonotic transmission of the virus may have happened in this area.

This is a putative but not plausible explanation.

Continual emigration (along with the use of unsafe syringes) between countries escalated the transmission of the virus.

Other scientists believe that the HIV-1C strain circulated in South America at around the same time that the HIV-1C strain was introduced in Africa. Very little research has been done on this theory because it is fairly young.

Transmission from non-humans to humans

The majority of HIV researchers agree that HIV evolved at some point from the closely related Simian immunodeficiency virus (SIV), and that SIV or HIV (post mutation) was transferred from primates to humans in the recent past.

History of HIV/AIDS

However, some loose ends remain unresolved.

It is not yet explained why only four HIV groups (HIV-1 groups M and O, and HIV-2 groups A and B) spread considerably in human populations, despite bushmeat practices being very widespread in Central and West Africa.

It also remains unexplained why all epidemic HIV groups emerged in humans nearly simultaneously, and only in the 20th century, despite very old human exposure to SIV. A recent phylogenetic study demonstrated that SIV is at least tens of thousands of years old.

The discovery of the main HIV / SIV phylogenetic relationships permits explaining broadly HIV biogeography. The early centers of the HIV-1 groups were in Central Africa, where the primate reservoirs of the related SIVcpz (chimpanzees) and SIVgor viruses (gorillas) exist.

Similarly, the HIV-2 groups had their centers in West Africa, where sooty mangabeys, which harbor the related SIVsmm virus, exist. However these relationships do not explain more detailed patterns of biogeography, such as why epidemic HIV-2 groups (A and B) only evolved in the Ivory Coast, which is one of only six countries harboring the sooty mangabey.

It is also unclear why the SIV naturally endemic in the chimpanzee subspecies (inhabiting the Democratic Republic of Congo, Central African Republic, Rwanda, Burundi, Uganda, and Tanzania) did not spawn an epidemic HIV-1 strain to natives of all these African countries, while the Democratic Republic of Congo was the main center of HIV-1 group M subspecies reservoir.

It seems that the mass injection (covert operation) with unsterilized needles rather than bush meat is the most scientifically valid contribution.

It is clear that the several HIV-1 and HIV-2 strains descend from SIVcpz, SIVgor, and SIVsmm viruses, and that bushmeat practice is not the most plausible avenue for cross species transfer to humans.

Origin and the Epidemic

Several of the theories of HIV origin put forward attempt to explain the unresolved puzzle.

Most of them accept the established knowledge of the HIV/SIV phylogenetic relationships, and also accept that bushmeat practice was the most likely cause of the initial transfer to humans, before Western colonization.

All of them propose that the simultaneous epidemic emergences of four HIV groups in the late 19th and early 20th century, and the lack of previous known emergences, are explained by new factor/s that appeared in the relevant African regions in that time frame. These new factor/s would have acted either to increase human exposures to SIV, to help it to adapt to the human organism by mutation (thus enhancing its humans transmission), or to cause an initial burst of transmissions crossing an epidemiological threshold, and therefore increasing the probability of continued spread.

Genetic studies of the virus suggested in 2008 that the most recent common ancestor of the HIV-1 M group dates back to the Belgian Congo city of Léopoldville (modern Kinshasa), 1910.

Proponents of this dating link the HIV epidemic with the emergence of colonialism and growth of large colonial African cities, leading to social changes, including a higher degree of sexual promiscuity, the spread of prostitution, and the concomitant high frequency of genital ulcer diseases (such as syphilis) in nascent colonial cities.

In 2014, a study conducted by scientists from the University of Oxford and the University of Leuven, in Belgium, revealed that because approximately one million people every year would flow through the prominent city of Kinshasa, which served as the origin of the first known HIV cases in the 1920s, passengers riding on the region's Belgian railway trains were able to spread the virus to larger areas. The study also attributed a roaring sex trade, rapid population growth and unsterilized needles used in health clinics as other factors which contributed to the emergence of the Africa HIV epidemic.

Colonialism in Africa

Amit Chitnis, Diana Rawls, and Jim Moore proposed that HIV may have emerged epidemically as a result of the harsh conditions, forced labor, displacement, and unsafe injection and vaccination practices associated with colonialism, particularly in French Equatorial Africa. The workers in plantations, construction projects, and other colonial enterprises were supplied with bushmeat, which would have contributed to an increase in hunting and, it follows, a higher incidence of human exposure to SIV. Several historical sources support the view that bushmeat hunting indeed increased, both because of the necessity to supply workers and because firearms became more widely available.

The colonial authorities also gave many vaccinations against smallpox, and injections, of which many would be made without sterilizing the equipment between uses (unsafe injections).

In addition, they proposed that the conditions of extreme stress associated with forced labor could depress the immune system of workers, therefore prolonging the primary acute infection period of someone newly infected by SIV, thus increasing the odds of both adaptation of the virus to humans, and of further transmissions.

The authors proposed that HIV-1 originated in the area of French Equatorial Africa in the early 20th century (when the colonial abuses and forced labor were at their peak).

Later researches proved these predictions mostly correct.

And all groups of HIV-1 descend from either SIVcpz or SIVgor from apes living to the west of the Ubangi River, either in countries that belonged to the French Equatorial African Federation of colonies, in Equatorial Guinea (then a Spanish colony), or in Cameroon (which was a German colony between 1884 and 1916, and then fell to Allied forces in World War I, and had most of its area administered by France, in close association with French Equatorial Africa).

Unsterile injections

In several articles published since 2001, Preston Marx, Philip Alcabes, and Ernest Drucker proposed that HIV emerged because of rapid serial human-to-human transmission of SIV (after a bushmeat hunter or handler became SIV-infected) through unsterilized injections.

The injection campaigns against sleeping sickness

David Gisselquist proposed that the mass injection campaigns to treat trypanosomiasis (sleeping sickness) in Central Africa were responsible for the emergence of HIV-1. Unlike Marx et al., Gisselquist argued that the millions of unsafe injections administered during these campaigns were sufficient to spread rare HIV infections into an epidemic, and that evolution of HIV through serial passage was not essential to the emergence of the HIV epidemic in the 20th century.

This theory focuses on injection campaigns that peaked in the period 1910–40, that is, around the time the HIV-1 groups started to spread. It also focuses on the fact that many of the injections in these campaigns were intravenous (which are more likely to transmit SIV/HIV than subcutaneous or intramuscular injections), and many of the patients received many (often more than 10) injections per year, therefore increasing the odds of SIV serial passage.

The authors suggested that the very high prevalence of the Hepatitis C virus in southern Cameroon and forested areas of French Equatorial Africa (around 40–50%) can be better explained by the unsterilized injections used to treat yaws, because this disease was much more prevalent than syphilis, trypanosomiasis, and leprosy in these areas. They suggested that all these parenteral risks caused, not only the massive spread of Hepatitis C but also the spread of other pathogens, and the emergence of HIV-1.

The same procedures could have exponentially amplified HIV-1, from a single hunter/cook occupationally infected with SIV to several thousand patients treated with arsenicals or other drugs, a threshold beyond which sexual transmission could prosper.

The Origins of AIDS, the virus can be traced to a central African bush hunter in 1921, with colonial medical campaigns using improperly sterilized syringe and needles playing a key role in. Pépin concludes that AIDS spread silently in Africa for decades, fueled by urbanization and prostitution since the initial cross-species infection. Pépin also claims that the virus was brought to the Americas by a Haitian teacher returning home from Zaire in the 1960s. Sex tourism and contaminated blood transfusion centers ultimately propelled AIDS to public consciousness in the 1980s and a worldwide pandemic.

Chapter 19

UFOs

UFOs have been subject of investigation for many years. Its scope and scientific analysis varied from its very first sighting known. There were peak periods and lull periods with many committees in between, headed by leading scientists and interested individuals. There was polarization of interest from enthusiasts to debunker to "won't care less souls".

I do not fall into any camp but wish to remain neutral since I do not have incontrovertible information, to go by.

There was abundance of misinformation.

It eventually, led to speculation and conspiracy theories. That is why I think, I should discuss it with some length, not because I do not have any other conspiracy material write about but it finally led to Disclosure Project which is currently live.

There is enough of other of conspiracies.

I have included only a few in this book.

I accidentally stumble upon, the UFO conspiracy, just as well. Governments or independent academics in the United States, Canada, the United Kingdom, Japan, Peru, France, Belgium, Sweden, Brazil, Chile, Uruguay, Mexico, Spain, and the Soviet Union are known to have investigated UFO reports at various times. Among the best known government studies are the ghost rockets investigation by the Swedish military (1946–1947), Project Blue Book, previously Project Sign and Project Grudge, conducted by the USAF from 1947 until 1969, the secret U.S.

Army/Air Force Project Twinkle investigation into green fireballs (1948–1951), the secret USAF Project Blue Book Special Report No. 14 by the Battelle Memorial Institute, and the Brazilian Air Force's 1977 Operação Prato (Operation Saucer). France has had an ongoing investigation (GEPAN/SEPRA/GEIPAN) within its space agency Centre National d'études Spatiales (CNES) since 1977. The government of Uruguay has had a similar investigation since 1989.

An unidentified flying object, or UFO, in its most general definition, is any apparent anomaly in the sky that is not identifiable as a known object or phenomenon. Culturally, UFOs were associated with claims of visitation by extraterrestrial life or government related conspiracy theories, and have become popular subjects in films and fiction. Some UFOs were later identified, sometimes identifications were not possible owing to the low quality of evidence related to UFO sightings (generally anecdotal evidence and eyewitness accounts) or due to callous disregard for proper scientific investigation or generally due to reporting them as hoax. Stories of fantastical celestial apparitions have been told since antiquity, but the term "UFO" (or "UFOB") was officially created in 1953 by the United States Air Force (USAF) to serve as a catch all "term" for all such sighting and reports. In its initial definition, the USAF stated that a "UFOB" was "any airborne object which by performance, aerodynamic characteristics, or unusual features, does not conform to any presently known aircraft or missile type, or which cannot be positively identified as a familiar object."

Accordingly, the term was initially restricted to that fraction of cases which remained unidentified after investigation, as the USAF was interested in the potential national security issues and/or "technical aspects" of such findings. During the late 1940s and through the 1950s, UFOs were often referred to (popularly) as "flying saucers" or "flying discs". The term UFO became more widespread during the 1950s, at first in technical literature, but later its use in popular journals. UFOs garnered considerable interest during the Cold War, an era associated with a heightened concern for national security. Various studies have concluded that the phenomenon does not represent neither a threat to national security nor does it contain anything worthy of scientific pursuit (e.g., 1951 Flying Saucer Working Party, 1953 CIA Robertson Panel, USAF Project Blue Book, Condon Committee).

Project Sign

Project Sign in 1948 produced a highly classified finding that the best UFO reports probably had an extraterrestrial explanation. A top secret Swedish military opinion given to the USAF in 1948 stated that some of their analysts believed that the 1946 ghost rockets and later flying saucers had extraterrestrial origins. In 1954 German rocket scientist Hermann Oberth revealed that an internal West German government investigation, which he headed, had arrived at an extraterrestrial conclusion, but this study was never made public.

Project Grudge

Project Sign was dismantled and became Project Grudge at the end of 1948. Angered by the low quality of investigations by Grudge, the Air Force Director of Intelligence reorganized it as Project Blue Book in late 1951, placing Ruppelt in charge. Blue Book closed down in 1970, using the Condon Committee's negative conclusion as a rationale, thus ending official Air Force investigation into UFOs.

However, a 1969 USAF document, known as the Bolender memo, along with later government documents, revealed that non-public U.S. government UFO investigations continued after 1970. The Bolender memo first stated that "reports of unidentified flying objects that could affect national security are not part of the Blue Book system," indicating that more serious UFO incidents were handled outside the public domain and not included in the Blue Book investigation.

The memo then added, "reports of UFOs which could affect national security would continue to be handled through the standard Air Force procedures designed for this purpose."

In addition, in the late 1960s a chapter on UFOs in the Space Sciences course at the U.S. Air Force Academy gave serious consideration to possible extraterrestrial origins. When this knowledge became public, the Air Force in 1970 issued a statement to the effect that the book was outdated and that the cadets instead were being informed of the Condon Report's negative conclusion.

Condon Committee

The Condon Committee was the informal name of the University of Colorado UFO Project, a group funded by the United States Air Force from 1966 to 1968 at the University of Colorado to study unidentified flying objects under the direction of physicist Edward Condon. The result of its work, formally titled Scientific Study of Unidentified Flying Objects, and known as the Condon Report, appeared in 1968.

After examining hundreds of UFO files from the Air Force's Project Blue Book, from the civilian UFO groups, National Investigations Committee On Aerial Phenomena (NICAP) and Aerial Phenomena Research Organization (APRO), and sightings reported and investigated during the life of the Project, the Committee produced a Final Report that said the study of UFOs was unlikely to yield major scientific discoveries.

The Report's conclusions were generally welcomed by the scientific community and have been cited as a decisive factor in the subsequent low level of interest in UFO activity among academics.

According to some it is "the most influential public document" concerning the scientific status of the UFO phenomenon. The authorities concerned strangely made a subtle directive as to, "all future or current scientific work on the UFO problem must make reference to the Condon Report".

I am not sure why such a directive was officially made, if not for the reason of keeping "a tight lid" on the UFO information.

USAF Regulation 200-2

Air Force Regulation 200-2, issued in 1953 and 1954, defined an Unidentified Flying Object ("UFOB") as "any airborne object which by performance, aerodynamic characteristics, or unusual features, does not conform to any presently known aircraft or missile type, or which cannot be positively identified as a familiar object." The regulation also said UFOBs were to be investigated as a "possible threat to the security of the United States" and "to determine technical aspects involved."

The regulation went on to say that "it is permissible to inform news media representatives on UFOB's when the object is positively identified as a familiar object," but added:

"For those objects which are not explainable, the ATIC (Air Technical Intelligence Center) will analyze only that fact and concur whether the data is worthy of release to the public domain, due to many unknowns factors involved in the investigations."

Project Blue Book

Project Blue Book was one of a series of systematic studies of unidentified flying objects (UFOs) conducted by the United States Air Force. It started in 1952, and it was the third study of its kind (the first, the Project Sign in 1947 and the second the Project Grudge in 1949). A termination order was given for the study in December 1969, and all activity under its auspices ceased in January 1970.

Project Blue Book had two goals;

To determine if UFOs were a threat to national security, and to scientifically analyze UFO related data.

Thousands of UFO reports were collected, analyzed and filed. As the result of the Condon Report (1968), which concluded there was nothing anomalous about UFOs, Project Blue Book was shut down in December 1969 and the Air Force continues to provide the following summary of its investigations;

No UFO reported, investigated and evaluated by the Air Force was ever an indication of threat to our national security;

There was no evidence submitted to or discovered by the Air Force that sightings categorized as "unidentified" represented technological developments or principles beyond the range of modern scientific knowledge; and there was no evidence indicating that sightings categorized as "unidentified" were extraterrestrial vehicles.

By the time Project Blue Book ended, it had collected 12,618 UFO reports, and concluded that most of them were misidentification of natural phenomena (clouds, stars, etc.) or conventional aircraft. According to the National Reconnaissance Office a number of the reports could be explained by flights of the formerly secret reconnaissance planes U-2 and A-12. A small percentage of UFO reports were classified as unexplained, even after stringent analysis. The UFO reports were archived and are available under the Freedom of Information Act, but names and other personal information of all witnesses have been redacted.

CUFOS

J. Allen Hynek, a trained astronomer who served as a scientific advisor for Project Blue Book, was initially skeptical of UFO reports, but eventually came to the conclusion that many of them could not be satisfactorily explained and was highly critical of what he described as "the cavalier disregard by Project Blue Book of the principles of scientific investigation." Leaving government work, he founded the privately funded CUFOS, to whose work he devoted, the rest of his life. Other private groups studying the phenomenon include the MUFON, a grass root organization whose investigator's handbooks go into great detail on the documentation of alleged UFO sightings. Like Hynek, Jacques Vallée, a scientist and a prominent UFO researcher, has pointed to what he believes is the scientific deficiency of most UFO research, including government studies.

He complains of the mythology and cultism often associated with the phenomenon, but alleges that several hundred professional scientists—a group both he and Hynek have termed "the invisible college"—will continue to study UFOs in private.

MUFON

The Mutual UFO Network (MUFON) is an American-based non-profit organization that investigates cases of alleged UFO sightings. It is one of the oldest and largest civilian UFO investigative organizations in the United States. MUFON was originally established as the Midwest UFO Network in Quincy,

Illinois on May 31st, 1969 by Walter H. Andrus, Allen Utke, John Schuessler, and others. Most of MUFON's early members had earlier been associated with the Aerial Phenomena Research Organization (APRO).

The organization claims to have more than 3,000 members worldwide, with a majority of its membership base situated in the continental United States. MUFON operates a worldwide network of regional directors, holds an annual international symposium, and publishes the monthly MUFON UFO Journal. The group now has more than 75 field investigators, as well as specialized teams to investigate possible physical evidence of any extraterrestrial craft. The network trains volunteers to be investigators, and teaches them how to interview witnesses, perform research, and how to draw conclusions from the evidence.

Although investigators are not paid, they must pass both an examination based on a 265 page manual, and a background check. The stated mission of MUFON is the study of UFOs for the benefit of humanity through investigations, research and education.

Along with the J. Allen Hynek Center for UFO Studies (CUFOS) and the Fund for UFO Research (FUFOR), MUFON is part of the UFO Research Coalition, a collaborative effort by the three main UFO investigative organizations in the US whose goal is to share personnel and other research resources, and to fund and promote the scientific study of the UFO phenomenon. MUFON is currently headquartered in Newport Beach, California under the direction of Jan Harzan.

The Soviet UFO Sightings

Unidentified Flying Objects UFOs - have been sighted over the territory we know today as Russia since the days of antiquity. For centuries people have seen objects in the sky that they could not identify, and many of them have recorded sightings, which could not be explained away as meteors, planets, stars, or weather balloons. Some of the most interesting information concerning UFO sightings is still locked away in the secret archives of the state. Occasionally, the guarded vaults do open up, either by the passing of history or by chance, and information leaks out. Recently declassified documents of the Russian Ministry of the Interior, dating back to the beginning of the 19th century, revealed some interesting UFO sightings from the Russian Empire. Among them was a very unusual report to the Tsar from his Third Department of the Chancellery (the former title of the secret police). The report describes certain extraordinary light effects observed in the sky by the inhabitants of the city of Orenburg, and corroborated by the police and military, during the night of December 26th,1830. Other reports mentioned the appearance of UFOs over Ustyug on January 30, 1844, as well as sightings from 1846 and 1847.

St. Petersburg UFO

A giant bright, spherical object flew over St. Petersburg on July 30th, 1880. The UFO was accompanied by two identical crafts, only smaller in size. The flight of the UFOs was noiseless, and they were observed over the city for three minutes.

UFOs, Area 51 and Conspiracy

Its secretive nature and undoubted connection to classified aircraft research, together with reports of unusual phenomena, have led Area 51 to become a focus of modern UFO conspiracy theories. Some of the activities mentioned in such theories at Area 51 include:

The storage, examination, and reverse engineering of crashed alien spacecraft (including material supposedly recovered at Roswell), the study of their occupants (living and dead), and the manufacture of aircraft based on alien technology.

Meetings or joint undertakings with extraterrestrials.

The development of exotic energy weapons for the Strategic Defense Initiative (SDI) or other weapon programs.

The development of means of weather control.

The development of time travel and teleportation technology.

The development of unusual and exotic propulsion systems related to the Aurora Program.

Activities related to a supposed shadowy one world government or the Majestic 12 organization.

Many of the hypotheses concern the underground facilities at Groom or at Papoose Lake (also known as "S-4 location") and include claims of a transcontinental underground railroad system, a disappearing airstrip (nicknamed the "Cheshire Airstrip", after Lewis Carroll's Cheshire cat) which briefly appears when water is sprayed onto its camouflaged asphalt, and engineering based on alien technology.

Publicly available satellite imagery, however, reveals clearly visible landing strips at Groom Dry Lake, but not at Papoose Lake.

In the mid 1950s, civilian aircraft flew under 20,000 feet while military aircraft flew under 40,000 feet. Once the U-2 began flying at above 60,000 feet, an unexpected side effect was an increasing number of UFO sighting reports. Sightings occurred most often during early evenings hours, when airline pilots flying west saw the U-2's silver wings reflect the setting sun, giving the aircraft a "fiery" appearance. Many sighting reports came to the Air Force's Project Blue Book, which investigated UFO sightings, through air traffic controllers and letters to the government. The project checked U-2 and later OXCART flight records to eliminate the majority of UFO reports it received during the late 1950s and 1960s.

However, it could not or did not reveal "the truth behind what they saw" to the letter writers.

Why this was not explained then, is a mystery to many.

Center for the Study of Extraterrestrial Intelligence

The Center for the Study of Extraterrestrial Intelligence (more commonly known as CSETI) is a worldwide organization focusing on collecting information about UFOs, with a specific interest in extraterrestrial life forms.

It was founded as a non-profit organization by Dr. Steven Greer, who has been the head of the organization since its inception in 1990, with the stated aim of "establishing peaceful and sustainable relations with extraterrestrial life forms".

The official statements regarding its intentions also included a new category of extraterrestrial encounters, namely CE-5 or "close encounters of the fifth kind". This was defined by Greer as human initiated contact and/or communication with extraterrestrial life.

Since its inception, the organization has spent anywhere between $3.5 million to $5 million to achieve its goals.

Though most of its claims have been rejected after outside scrutiny, the organization claims to have over 3,000 "confirmed" reports of UFO sightings by pilots, and over 4,000 proofs of what they describe as 'landing traces'. The latter refers to incidences where UFOs supposedly have left behind trace evidence, such as electromagnetic readings, at landing sites on Earth.

The organization utilizes;

"Rapid Mobilization Investigative Teams" with the aim of arriving at landing sites as quickly as possible to remove any evidence that may be available.

In April 1997, the organization made representation to US Members of Congress and placed "their collected evidence" for examination along with putative theories surrounding UFOs and extraterrestrial visitations.

They were aided in this endeavor by Apollo 14 astronaut Edgar Mitchell. After the initial briefing, Greer and CSETI demanded a full hearing regarding their supposed evidence, as it would allow them to subpoena witnesses, and protect the confidentiality of witnesses who would otherwise not come forward.

Congress did not grant them the hearing they requested.

Greer also states he was allowed an audience with James Woolsey, former director of the CIA, although Woolsey described it as a dinner party at which he politely listened to Greer.

A subpoena is a writ issued by a government agency, most often a court, to compel testimony by a witness or production of evidence under a penalty for failure.

There are two common types of subpoena:

subpoena ad testificandum orders a person to testify before the ordering authority or face punishment. The subpoena can also request the testimony to be given by phone or in person.

subpoena duces tecum orders a person or organization to bring physical evidence before the ordering authority or face punishment. This is often used for requests to mail copies of documents to the requesting party or directly to the court.

Disclosure

In the early 2000s, the concept of "disclosure" became increasingly popular in the UFO enthusiasts: The phrase, "UFO conspiracy" was added to the vocabulary to emphasize that the government had suppressed information on alien contact and full disclosure was needed.

This was pursued actively by lobbyists.

In 1993, Steven M. Greer founded the Disclosure Project to promote the concept. In May 2001, Greer held a press conference at the National Press Club in D.C that demanded Congress hold hearings on "Secret U.S. involvement with UFOs and extraterrestrials".

It was described by an attending BBC reporter as "the strangest ever news conference hosted by Washington's August National Press Club."

The Disclosure Project's claims were met with ridicule, derision and skepticism by authorized spokespeople of the U. S. Air Force. Subsequently, Stephen G. Bassett's Paradigm Research Group held a "Citizen Hearing on Disclosure" at the National Press Club from 29th April to 3rd May 2013. The group paid former U.S. Senators Mike Gravel, Carolyn Cheeks, Kilpatrick, Roscoe Bartlett, Merrill Cook, Darlene Hooley, and Lynn Woolsey, $20,000 each for there participation.

One of the conveners, Kilpatrick openly complained about the lack of transparency by those representing the state, at the committee meeting. Other such groups include Citizens Against UFO Secrecy, founded in 1977.

Brian Todd O'Leary

Brian Todd O'Leary (January 27th, 1940 – July 28th, 2011) was an American scientist, author, and former NASA astronaut. He was a member of the sixth group of astronauts selected by NASA in August 1967. The members of this group of eleven were known as the scientist-astronauts, intended to train for the Apollo Applications Program — a follow-on to the Apollo program, which was abruptly canceled. O'Leary was born and raised in Boston, Massachusetts on January 27th, 1940. He admitted that his teenage visit to Washington, D.C. inspired him in exploration of uncharted territories. Armed with patriotism he drove his efforts to become an astronaut.

Climbing the Matterhorn, running the Boston Marathon and becoming an Eagle Scout in the Boy Scouts of America were among many activities, that made him build up the stamina necessary for a risky Apollo Mission.

He enjoyed photography, hiking, cartooning, jazz piano and yoga.

O'Leary graduated from Belmont High School, Belmont, Massachusetts in 1957. He received a Bachelor's Degree in Physics from Williams College in 1961, a Master Degree in Astronomy from Georgetown University in 1964 and a Doctor of Philosophy in Astronomy from the University of California, Berkeley in 1967.

O'Leary was chairman of the board of directors of the Institute for Security and Cooperation in Space; 1990, founding board member of the International Association for New Science;

2003, founding president of the New Energy Movement; 2007–2011

While attending graduate school in astronomy at the University of California, Berkeley, O'Leary published several scientific papers on the atmosphere of Mars. O'Leary's Ph.D. thesis in 1967 was on the Martian surface. Soon after completing his Ph.D. thesis, O'Leary was the first astronaut specifically selected for a potential manned Mars mission when it was still in NASA's program plan, projected for the 1980s as a follow-on to the Apollo lunar program. O'Leary was the only planetary scientist-astronaut in NASA Astronaut Corps during the Apollo program.

O'Leary resigned from the astronaut program in April 1968, and cited several reasons for resigning in his book "The Making of an Ex-Astronaut".

After O'Leary's resignation from NASA, Carl Sagan recruited him to teach at Cornell University in 1968, where he researched and lectured until 1971 as a research associate (1968-1969) and assistant professor of astronomy (1969-1971). While at Cornell, he studied lunar mascons. During the 1970-1971 academic year, O'Leary was deputy team leader of the Mariner 10 Venus-Mercury TV Science Team as a visiting researcher at the California Institute of Technology. The team received NASA's group achievement award for its participation.

He subsequently taught astronomy, physics, and science policy assessment at several academic institutions, including San Francisco State University (associate professor of astronomy and interdisciplinary sciences; 1971-1972), the UC Berkeley School of Law (visiting associate professor; 1971–1972), Hampshire College (assistant professor of astronomy and science policy assessment; 1972–1975), Princeton University (research staff and lecturer in physics; 1976–1981) and California State University, Long Beach (visiting lecturer in physics; 1986-1987).

O'Leary authored several popular books and more than one hundred peer reviewed articles in the fields of planetary science, astronautics, and science policy.

During those years, he also immersed himself in several controversies relating to NASA's objectives, including its manned lunar landings, the Space Shuttle, and the weaponization of space. He promoted a joint manned mission to Mars between the U.S. and the Soviet Union. O'Leary twice traveled to the Soviet Union in the late 1980s to promote the peaceful exploration of space.

A remote viewing experience in 1979 and a near-death experience in 1982 initiated O'Leary's departure from orthodox science. After Princeton, O'Leary worked in the space industry at Science Applications International Corporation in Hermosa Beach, California, beginning in 1982. He refused to work on military space applications, for which reason he lost his position there in 1987.

Beginning in 1987, O'Leary increasingly explored unorthodox ideas, particularly the relationship between consciousness and science, and became widely known for his writings on "the frontiers of science, space, energy and culture".

He advocated scientific testing of phenomena not currently recognized by orthodox science.

In the mid 1990s, he began to write about his investigations regarding innovative technologies that allegedly utilize energy sources that science does not currently recognize (also called "new energy solutions"), and how those technologies can transform the planet and the human journey.

He believed there is an extraterrestrial presence on earth, and their relationship to innovative technologies that could potentially transform earth. He also believed in conjoined organized suppression of the presence of extraterrestrials and their technology.

In 2003, O'Leary founded the New Energy Movement. Shortly after his new energy colleague Eugene Mallove was murdered in 2004, he moved to Ecuador, where he resided for the rest of his life. He continued to travel and publicly lecture on the subject of new energy and planetary healing.

In 2005, O'Leary wrote the foreword to Steven M. Greer's Hidden Truth, Forbidden Knowledge, which is concerned with the extraterrestrial presence on Earth and related free energy, anti-gravity and other exotic technologies.

Steven Macon Greer

Steven Macon Greer (June 28th, 1955) is an American retired medical doctor and ufologist who founded the Center for the Study of Extraterrestrial Intelligence (CSETI) and The Disclosure Project, which seeks the disclosure of allegedly suppressed UFO information.

Greer was born in Charlotte, North Carolina in 1955. Greer claims to have seen an unidentified flying object at close range when he was about eight years old, which inspired his interest in ufology.

He was trained as a Transcendental Meditation teacher and served as the director of the meditation organization. It is interesting to note that he was a meditation teacher before he entered the Medical Faculty.

Greer completed his graduate work at East Tennessee State University James H. Quillen College of Medicine in 1987. He attended MAHEC University of North Carolina where he completed his internship in 1988 and received his Virginia medical license in 1989. That year he became a member of the Alpha Omega Alpha Honor Medical Society.

In 2001, CSETI hosted a press conference in Washington, D.C., regarding their evidence and collected statements, which contained information from the former Investigation branch of the US Federal Aviation Authority, John Callahan.

Callahan also claims to have met with the C.I.A., then under the Reagan administration, though in this instance the C.I.A., has emphatically denied the meeting took place.

John Callahan

The following is an excerpt taken from over 35 hours of witness testimony in the CSETI archives.

For 6 years Mr. Callahan was the Division Chief of the Accidents and Investigations Branch of the FAA in Washington DC. In his testimony he tells about a 1986 Japanese Airlines 747 flight that was followed by a UFO for 31 minutes over the Alaskan skies. The UFO also trailed a United Airlines flight until the flight landed. There was visual confirmation as well as air-based and ground-based radar confirmation. This event was significant enough for the then FAA Administrator, Admiral Engen, to hold a briefing where the FBI, CIA, President Reagan's Scientific Study Team, as well as others attended. Videotape radar evidence, air traffic voice communications and paper reports were compiled and presented. At the conclusion of this meeting, the attending CIA members instructed everyone present that "the meeting never took place" and that "this incident was never recorded." Not realizing that there was additional evidence, they confiscated just the evidence presented, but Mr. Callahan was able to secure videotape and audio evidence of the event.

"What I can tell you is what I've seen with my own eyes. I've got a videotape. I've got the voice tape. I've got the reports that were filed that will confirm what I've been telling you. And I'm one of those, what you would call the high Government officials in the FAA".

"It still bothers me that I've seen all this, I know all this, and I'm walking around with the answer, and nobody wants to ask the question to get the answer. And it kind of irritates me a little bit. And I don't believe our Government should be set up that way. I think when we have something like this, that you can probably find out more about what's going on in the world (by not covering it up)".

The Anchorage Incident

Japan Air Lines flight 1628 was a UFO incident that occurred on November 17th, 1986 involving a Japanese Boeing 747 cargo aircraft. The aircraft was en route from Paris to Narita International Airport, near Tokyo, with a cargo of Beaujolais wine. On the Reykjavík to Anchorage section of the flight, at 5:11 PM over eastern Alaska, the crew first witnessed two unidentified objects to their left. These abruptly rose from below and closed in to escort their aircraft. Each had two rectangular arrays of what appeared to be glowing nozzles or thrusters, though their bodies remained obscured by darkness. When closest, the aircraft's cabin was lit up and the captain could feel their heat in his face. These two craft departed before a third, much larger disk shaped object started trailing them, causing the pilots to request a change of course.

Anchorage Air Traffic Control obliged and requested an oncoming United Airlines flight to confirm the unidentified traffic, but when it and a military craft sighted JAL 1628 at about 5:51 PM, no other craft could be distinguished.

The sighting of 31 minutes ended in the vicinity of Denali.

FAA Reporting

Because of their radar deficiencies, when pilots seeing an unusual flying object, the FAA will not investigate unless the object can be identified by an air born pilot. Instead, the FAA will offer a host of weak explanations for the unknown object. If the FAA cannot identify the object within FAA terminology, then it doesn't exist. Another cliché they sometimes use:

"For every problem there is a solution. If there is no solution, there is no problem".

The Alaskan UFO investigation is a case in point.

The final FAA report concluded that the radar returns from Anchorage were simply a "split image" due to a malfunction in the radar equipment, which showed occasional second blips that had been mistaken for the UFO.

Thus the FAA would not confirm that the incident took place. Yet all three controllers engaged with the pilot during the extended sighting filed statements that contradict this finding. "Several times I had single primary returns where L1628 reported traffic," wrote one.

"I observed data on the radar that coincided with information that the pilot of Jl1628 reported," stated another.

The FAA spokesman at the time, Paul Steucke, said it was just a "coincidence" that the split image happened to fall at the right distance and the same side of the aircraft where the object was reported visually by the pilot. And the final report simply ignored outright, the three visual sightings with all their details and drawings, as if the event had never happened.

Other Sightings

Michael Smith was an Air Traffic Controller with the Air Force in Oregon and, subsequently, in Michigan. At both of these facilities he and others witnessed UFOs tracked on radar and moving at extraordinary speeds. He also confirmed that personnel were expected to maintain secrecy concerning these observations, and that NORAD, the North American Air Defense Command, was fully apprised of these events.

Chuck Sorrells is a career Air Force military man who was at Edwards Air Force Base in 1965 when not one, but at least seven UFOs appeared over Edwards Air Force Base airspace, moving in extraordinary fashion at enormous speeds, making right hand turns and other maneuvers which no known aircraft was capable of at the time.

This event lasted for five or six hours.

Stephen G. Bassett

Stephen G. Bassett is the first, and only, extraterrestrial life (ET) lobbyist in the United States, treasurer of the political action committee Extraterrestrial Phenomena Political Action Committee and executive director of Paradigm Research Group (PRG) that says it wants to end the governments imposed truth embargo regarding the facts of ET's engaging with the human race.

Bassett's involvement with ET political activism began in 1996 when he worked as a volunteer for Harvard Medical

School's Pulitzer Prize-winning professor John Mack's Program for Extraordinary Experience Research.

Also in 1996, Bassett registered in Washington D.C.. the first ET issue lobbyist (which he remains the only one) and in 1999 set up his political action committee Extraterrestrial Phenomena Political Action Committee.

In 2002, Bassett ran as an independent for the United States Congress in Maryland's 8th congressional district on a pro-ET disclosure platform, but lost receiving only 1% of the vote.

In 2013, Bassett's Paradigm Research Group sponsored the April 29th to May 3rd Citizen Hearing on Disclosure held at Washington D.C.'s National Press Club where expert witnesses gave sworn testimony to 5 former US congressmen.

One of the senators, Kilpatrick openly complained about the lack of transparency by those who represented the state and alleged covered up of the facts regarding ETs.

John Lear

Son of the inventor of the Lear Jet, John Lear is one of the most controversial and important figures surrounding UFO conspiracy theories. He is a retired airline captain, multiple world record holder, and the only pilot to hold every FAA airplane certificate. He has flown numerous missions worldwide for the CIA and other government agencies.

He is also one of the core members of Pilots for 9/11 Truth.

John Lear is quite a character and the information that he puts out to the public ranges from underground and alien Moon bases to a planet where soul farming is done by extraterrestrial entities. He is one of the key figures in the Bob Lazar incident and also involved with the prominent individuals surrounding the infamous MJ-12 and "Avery" affairs.

In addition, Mr. Lear is perhaps one of the more notable individuals—besides Richard C. Hoagland—who believes in a secret space program that is far more advanced and superior than the space program that the general public is familiar with and is allowed to know about it.

Richard C. Hoagland

Richard Charles Hoagland (born April 25th, 1945), is an American author, and a proponent of various conspiracy theories about NASA, lost alien civilizations on the Moon and on Mars.

His writings claim that advanced civilizations exist or once existed on the Moon, Mars and on some of the moons of Jupiter and Saturn, and that NASA and the United States government have conspired to keep these facts secret.

He has advocated his ideas in two published books, videos, lectures, interviews, and press conferences.

Hoagland has been described by James Oberg of The Space Review, Phil Plait of Badastronomy.com, and Ralph Greenberg, a professor at Washington University, as a conspiracy theorist and fringe pseudo-scientist.

His book publisher describes him as "a unique mixture of amateur scientist, genius inventor, scam artist, and performer, blending true, legitimate, speculative science with his own extrapolations, tall tales, and inflations."

Milton William "Bill" Cooper

Milton William "Bill" Cooper (May 6th, 1943 – November 6, 2001) was an American conspiracy theorist, radio broadcaster, and author best known for his 1991 book Behold a Pale Horse, in which he warned of multiple global conspiracies, some involving extraterrestrial aliens. Cooper also described HIV/AIDS as a man-made disease used to target blacks, Hispanics, and homosexuals, and that a cure was made before it was implemented.

UFOs, Aliens and the Illuminati

Cooper caused a sensation in UFO circles in 1988 when he claimed to have seen secret documents while in the Navy describing governmental dealings with extraterrestrial aliens, a topic on which he expanded in Behold a Pale Horse. By one account he served as a "low level clerk" in the Navy, and as such would not have had the security clearance needed to access classified documents. Ufologists later asserted that some of the materials that Cooper claimed to have seen in Naval Intelligence documents were actually plagiarized verbatim of research by ufologists and were fabricated.

Don Ecker of UFO Magazine ran a series of exposes on Cooper in 1990.

William Cooper is best known for this 1991 book Behold a Pale Horse, which discusses various aspects of the New World Order and contains his many claims to have seen classified government documents while in the Navy.

The title is from the Bible's Book of Revelation.

"And I looked, and behold a pale horse: and his name that sat on him was Death, and Hell followed with him."

In the forward to Behold a Pale Horse, Cooper recounts his time in the Navy and begins by explaining that he and his fellow ship mates had seen several UFOs rise up from the ocean and disappear into the clouds. He insists that the Navy's sonar had also detected these objects. He explains that a superior officer then made him sign an agreement that he was not to discuss the event with anyone, including the other witnesses.

Cooper linked the Illuminati with his beliefs that extraterrestrials were secretly involved with the United States government, but later retracted these claims. He accused Dwight D. Eisenhower of negotiating a treaty with extraterrestrials in 1954, then establishing an inner circle of Illuminati to manage relations with them and keep their presence a secret from the general public. Cooper believed that aliens "manipulated and/or ruled the human race through various secret societies, religions, magic, witchcraft, and the occult", and that even the Illuminati were unknowingly being manipulated by them.

Cooper described the Illuminati as a secret international organization, controlled by the Bilderberg Group, that conspired with the Knights of Columbus, Masons, Skull and Bones, and other organizations. Its ultimate goal, he said, was the establishment of a New World Order. According to Cooper the Illuminati conspirators not only invented alien threats for their own gain, but actively conspired with extraterrestrials to take over the world.

Cooper believed that James Forrestal's fatal fall from a window on the sixteenth floor of Bethesda Hospital was connected to the alleged secret committee Majestic 12, and that JASON advisory group scientists reported to an elite group of Trilateral Commission and Council on Foreign Relations executive committee members who were high ranking members of the Illuminati.

Cooper seemed to be an honest and religious man. UFO sighting may have conflicted with his religious beliefs but all the same he had the courage to divulge this information, in his book but with some inconsistencies.

He feared the secret service, especially because he was coerced, not to leak classified information. Nevertheless, he began to leak information, often receiving threats to his life and his family. He continued undaunted, but has been increasingly harassed for his whistle-blowing.

Two weeks before his death, he decided he could not hold back what he had learned, all his life any longer, and blew the whistle on the whole charade, naming names, dates, places and everything else, he had known. Apparently this was the last straw. William Cooper's stunning presentation two weeks before his untimely death was bound to get him killed. He knew it, and stated it before his live audience. If anyone thinks this man's statements were false altogether, ask yourself "Why he was killed?"

There was tight security surrounding the details of the actual incident resulting in his death. The story was being spun to make it sound like he had a scuffle with two Sheriff deputies in which he shot one officer on the head twice. Then he was shot dead, according to the official story.

"The deputy is said to be alive, but Cooper was killed on the spot." You can bet your bottom dollar and nickels, that the truth surrounding this "assassination" will never be revealed once it is spun beyond recognition. There is little doubt that Bill simply went too far. This is the same sort of thing that got Kennedy shot.

If "they" kill presidents to cover up the tracks, how much chance does one lone ex-Navy officer have?

His expertise and knowledge, however, were of a military and political in nature (governmental, historical and cover ups), and not specifically scientific. The Associated Press published an article the day after his death labeling Cooper a "national leader of the militia movement" and tried to tie him to Timothy McVeigh saying "McVeigh, who was executed in May for the bombing of the federal building in Oklahoma City, listened to Cooper's broadcasts for inspiration, according to testimony by James Nichols, brother of Oklahoma bombing co-defendant Terry

Nichols during a 1996 pretrial hearing. While Cooper's death is a tragedy and it's slanderous to tie him to Timothy McVeigh, and many of the issues he discussed on his radio show and in his book are true, it is inexcusable that he fabricated tales of seeing "Top Secret" documents in an attempt to add credibility to his story and sell more books.

Stanton T. Friedman

Stanton Friedman (born July 29th, 1934) is a retired nuclear physicist and professional ufologist who resides in Fredericton, New Brunswick, Canada. He is the original civilian investigator of the Roswell incident. He worked on research and development projects for several large companies.

In 1970, Friedman left full time employment as a physicist to pursue the scientific investigation of UFOs. Since then, he has given lectures at more than 600 colleges and to more than 100 professional groups in 50 states, nine provinces, and 16 countries outside the USA. Additionally, he has worked as a consultant on the topic.

He has published more than 80 UFO-related papers and has appeared on many radio and television programs. He has also provided written testimony to Congressional hearings and appeared twice at the United Nations.

Friedman has consistently favoured use of the term "flying saucer" in his work, saying "Flying saucers are, by definition, unidentified flying objects, but very few unidentified flying objects are flying saucers. I am interested in the latter, not the former."

Friedman used to refer to himself as "The Flying Saucer Physicist", because of his degrees in nuclear physics and work on nuclear projects.

Friedman's positions regarding UFO phenomena

Friedman was the first civilian to document, the site of the Roswell UFO incident, and supports the hypothesis that it was a genuine crash of an extraterrestrial spacecraft. In 1968 Friedman told a committee of the U.S. House of Representatives that the evidence suggests that Earth is being visited by intelligently controlled extraterrestrial vehicles. Friedman also stated he believed that UFO sightings were consistent with crafts utilizing "magnetohydrodynamic" propulsion.

In 1996, after researching and fact checking the Majestic 12 documents, Friedman said that there was no substantive grounds for dismissing their authenticity.

In 2004, on George Noory's Coast to Coast radio show, Friedman debated Seth Shostak, the SETI Institutes Senior Astronomer. Like Friedman, Shostak also believes in the existence of intelligent life other than humans; however, unlike Friedman, he doesn't believe such life is now on Earth or the documents related to UFO sightings are true.

Friedman has hypothesized that UFOs may originate from relatively nearby sun like stars. A piece of evidence that he often cites with respect to this hypothesis is the 1964 star map drawn by alleged alien abductee Betty Hill during a hypnosis session, which she said was shown to her during her abduction.

Astronomer Marjorie Fish constructed a three dimensional map of nearby sun like stars and claimed a good match from the perspective of Zeta Reticuli, about 39 light years away.

Barney and Betty Hill

Barney and Betty Hill were an American couple who were allegedly abducted by extraterrestrials in a rural portion of New Hampshire from September 19th to September 20th, 1961.

The incident came to be called the "Hill Abduction" or the "Zeta Reticuli Incident" because the couple stated they had been kidnapped for a short time by a UFO. It was the first widely publicized report of alien abduction, adapted into the best-selling 1966 book "The Interrupted Journey" and the 1975 television movie The UFO Incident.

Most of Betty Hill's notes, tapes, and other items have been placed in the permanent collection at the University of New Hampshire, her alma mater. In July 2011, the state Division of Historical Resources marked the site of the alleged craft's first approach with a historical marker.

UFO encounter

According to a variety of reports given by the Hills, the alleged UFO sighting happened on September 19th, 1961, around 10:30 p.m. The Hills were driving back to Portsmouth from a vacation in Niagara Falls and Montreal. There were only a few other cars on the road. Just south of Lancaster, New Hampshire, Betty claimed to have observed a bright point of light in the sky that moved from below the moon and the planet Jupiter, upward to the west of the moon. While Barney navigated U.S. Route 3, Betty reasoned that she was observing a falling star, only it moved upward. Since it moved erratically and grew bigger and brighter, Betty urged Barney to stop the car for a closer look, as well as to walk their dog, Delsey. Barney stopped at a scenic picnic area just south of Twin Mountain. Worried about the presence of bears, Barney retrieved a pistol that he kept in the trunk of the car.

Betty, through binoculars, observed an "odd-shaped" craft flashing multi-colored lights that traveled across the face of the moon. Because her sister had confided to her about having a flying saucer sighting several years earlier, Betty thought it might be what she was observing.

What Barney observed, through binoculars, he reasoned out was a commercial airliner traveling toward Vermont on its way to Montreal. However, he soon changed his mind, because without "looking as if it had turned", the craft rapidly descended in his direction. This observation caused Barney to realize, "this object that was a plane was not a plane." He quickly returned to the car and drove toward Franconia Notch, a narrow, mountainous stretch of the road.

The Hills claimed that they continued driving on the isolated road, moving very slowly through Franconia Notch in order to observe the object as it came even closer. At one point, the object passed above a restaurant and signal tower on top of Cannon Mountain. It passed over the mountain and came out near the Old Man of the Mountain. Betty testified that it was at least one and a half times the length of the granite cliff profile, which was 40 feet (12 m) long, and that seemed to be rotating.

The couple watched as the silent, illuminated craft moved erratically and bounced back and forth in the night sky.

Approximately one mile south of Indian Head, they said, the object rapidly descended toward their vehicle causing Barney to stop directly in the middle of the highway. The huge, silent craft hovered approximately 80–100 feet (24–30 m) above the Hills' 1957 Chevrolet Bel Air and filled the entire field of the windshield.

It reminded Barney of a huge pancake.

Carrying his pistol in his pocket, he stepped away from the vehicle and moved closer to the object. Using the binoculars, Barney claimed to have seen about 8 to 11 humanoid figures who were peering out of the craft's windows, seeming to look at him. In unison, all but one figure moved to what appeared to be a panel on the rear wall of the hallway that encircled the front portion of the craft. The one remaining figure continued to look at Barney and communicated a message telling him to "stay where you are and keep looking."

Barney had a recollection of observing the humanoid forms wearing glossy black uniforms and black caps. Red lights on what appeared to be bat-wing fins began to telescope out of the sides of the craft and a long structure descended from the bottom of the craft. The silent craft approached to what Barney estimated was within 50–80 feet (15–24 m) overhead and 300 feet (91 m) away from him.

On October 21st, 1961, Barney reported to NICAP Investigator Walter Webb, that the "beings were somehow not human".

Barney "tore" the binoculars away from his eyes and ran back to his car. In a near hysterical state, he told Betty;

"They're going to capture us!"

He saw the object again shift its location to directly above the vehicle. He drove away at high speed, telling Betty to look for the object. She rolled down the window and looked up. Almost immediately, the Hills heard a rhythmic series of beeping or buzzing sounds which they said seemed to bounce off the trunk of their vehicle. The car vibrated and a tingling sensation passed through the Hills' bodies. The Hills said that at this point in time they experienced the onset of an altered state of consciousness that left their minds dulled. A second series of beeping or buzzing sounds returned the couple to full consciousness. They found that they had traveled nearly 35 miles (56 km) south but had only vague, spotty memories of this section of road. They recalled making a sudden unplanned turn, encountering a roadblock, and observing a fiery orb in the road.

Arriving home at about dawn, the Hills assert that they had some odd sensations and impulses they could not readily explain: Betty insisted their luggage be kept near the back door rather than in the main part of the house.

Their watches would never run again.

Barney said that the leather strap for the binoculars was torn, though he could not recall it tearing. The toes of his best dress shoes were scraped. Barney says he was compelled to examine his genitals in the bathroom, though he found nothing unusual. They took long showers to remove possible contamination and each drew a picture of what they had observed.

Perplexed, the Hills say they tried to reconstruct the chronology of events as they witnessed the UFO and drove home. But immediately after they heard the buzzing sounds, their memories became incomplete and fragmented. After sleeping for a few hours, Betty awoke and placed the shoes and clothing she had worn during the drive into her closet, observing that the dress was torn at the hem, zipper and lining. Later, when she retrieved the items from her closet, she noted a pinkish powder on her dress. She hung the dress on her clothesline and the pink powder blew away. But the dress was irreparably damaged. She threw it away, but then changed her mind, retrieving the dress and hanging it in her closet.

Over the years, five laboratories have conducted chemical and forensic analyses on the dress.

There were shiny, concentric circles on their car's trunk that had not been there the previous day. Betty and Barney experimented with a compass, noting that when they moved it close to the spots, the needle would whirl rapidly. But when they moved it a few inches away from the shiny spots, it would drop down.

The Star Map

In 1968, Marjorie Fish of Oak Harbor, Ohio read Fuller's Interrupted Journey. She was an elementary school teacher and amateur astronomer. Intrigued by the "star map", Fish wondered if it might be "deciphered" to determine which star system the UFO came from. Assuming that one of the fifteen stars on the map must represent the Earth's Sun, Fish constructed a three-dimensional model of nearby Sun-like stars using thread and beads, basing stellar distances on those published in the 1969 Gliese Star Catalogue. Studying thousands of vantage points over several years, the only one that seemed to match the Hill's map was from the viewpoint of the double star system of Zeta Reticuli.

Distance information needed to match three stars, forming the distinctive triangle Hill said she remembered, was not generally available until the 1969 Gliese Catalogue came out.

Fish sent her analysis to Webb. Agreeing with her conclusions, Webb sent the map to Terence Dickinson, editor of the popular magazine Astronomy. Dickinson did not endorse Fish and Webb's conclusions, but for the first time in the journal's history, Astronomy invited comments and debate on a UFO report, starting with an opening article in the December 1974 issue. For about a year afterward, the opinions page of Astronomy carried arguments for and against Fish's star map. Notable was an argument made by Carl Sagan and Steven Soter, arguing that the seeming "star map" was little more than a random alignment of chance points.

In contrast, those more favorable to the map, such as David Saunders, a statistician who had been on the Condon UFO study, argued that unusual alignment of key Sun-like stars in a plane centered around Zeta Reticuli (first described by Fish) was statistically improbable to have happened by chance from a random group of stars in our immediate neighborhood.

Development in 1980s

MJ-12

The so-called Majestic 12 documents surfaced in 1982, suggesting that there was secret, high-level U.S. government interest in UFOs dating to the 1940s.

Linda Moulton Howe

Linda Moulton Howe is an advocate of conspiracy theories that cattle mutilations are of extraterrestrial origin and speculations that the U.S. government is involved with aliens.

Milton William Cooper

In the 1980s, Milton William Cooper achieved a degree of prominence due to his conspiratorial writings.

Bob Lazar

In November 1989, Bob Lazar appeared in a special interview with investigative reporter George Knapp on Las Vegas TV station KLAS to discuss his alleged employment at S-4. In his interview with Knapp, Lazar said he first thought the saucers were secret, terrestrial aircraft, whose test flights must have been responsible for many UFO reports. Gradually, on closer examination and from having been shown multiple briefing documents, Lazar came to the conclusion that the discs must have been of extraterrestrial origin. In his filmed testimony, Lazar explains how this impression first hit him after he boarded the craft under study and examined their interior.

For the propulsion of the studied vehicles, Bob Lazar claims that the atomic Element 115 served as a nuclear fuel. Element 115 (provisionally named 'Ununpentium' (Uup)) reportedly provided an energy source which would produce anti-gravity effects under proton bombardment along with the production of antimatter used for energy production. Lazar's website says that, if the intense strong nuclear force field of element 115's nucleus was properly amplified, the resulting effect would be a distortion of the surrounding gravitational field, allowing the vehicle to immediately shorten the distance to a charted destination.

Lazar also claims that he was given introductory briefings describing the historical involvement with this planet for 10,000 years by extraterrestrial beings originating from the Zeta Reticuli 1 and 2 star system. These beings are therefore referred to as Zeta Reticulians, popularly called 'Greys'.

Lazar says he has degrees from the California Institute of Technology and Massachusetts Institute of Technology. In 1993, the Los Angeles Times looked into his background and found there was no evidence to support his claims.

USA / UFO Scenario

Because there so many overt and covert player/actors in the UFO and ET scenario, it is very difficult to unravel the relative truth or scientific basis.

Let me summarize few cardinal points.

They should not be taken on face value alone.

There is lot of red red herrings and misconceptions.

Number one which is obvious.

There are parallel governments in USA.

It could be minimum of three.

Those elected by popular vote by regular (rigid framework and term of office) elections. This, I call the changing or the flexible arm of the government. They actually do not have a clout in decision making.

The second is group made up of intellectuals coming from science, politics, defense affairs, foreign policy, trade and economy. They are either elected by their higher positioning in the respective fields or the power domains and influence they possess in business or politics. The number may amounts to as low six to twelve. They form the concrete policy group less amenable to change and provide continuity of governance irrespective of who heads the political mantle.

Both these groups on surface are democratic / republican institutes amenable to scrutiny albeit little. There is some structural rigidity alien to them however. Radical changes will never come about given this structural rigidity.

The third group and the most powerful is invincible and secretive and it has absolute power and wherewithal. The individuals in this group are not subjected to scrutiny since the above two institutions have abdicated their power to rein on this secretive society. The numbers in this group can be as little as five to innumerable. I believe this group has moles placed in the other two for eavesdropping and insinuating their will.

This leads to the rigidity of the governance.

Even the President has no power to intervene and if he/she trespasses, the group will stop, if necessary by assassination.

The other possibility of, this secretive group would train individuals as career serviceman / women is real. Nobody from the outside (independents) could get into this secretive system without heavy scrutiny.

Level of secrecy or classified information is the norm by default.

Evolution of disclosure is slow.

Freedom of information is almost non existent.

All these symptoms and signs of the lack of transparency in governance came into existence because of the people's interest on UFOs and ET/Aliens.

It does not matter whether there are aliens or not.

The American public has woken up from slumber after 70 years.

My entry in here, to US politics was accidental not coincidental. I always wished to have been to USA for work.

With above discovery, I do not think I have missed anything, much.

This form of rigid and hard to get information has led to many untoward outcomes. With the secret service operating with such stubbornness some of these guys/girls who look for truth/revelations, "give in" to pressure and "give up" without a fight.

The minority who persevere are frustrated to the core.

They in turn, shunt their frustration to form pressure groups. Often these pressure groups have few active members.

They are handicapped by lack of resources.

One has to use one's own pocket money.

The Internet has been very important to these pressure groups. Some of them are now organizing, even international symposiums.

There is a downside too.

Some of them go and publish books and some of these books are turned into films.

Facts often become fiction.

Some of them get rich but many become poor.

Often families break down and divorce ensues.

Initial enthusiasm wanes and the project comes to a standstill.

This is where one has to come forward and pump in money (donations) and jump start the stalled engine of inquiry.

These smart people do not wish to beg.

So it is imperative that when one is in the break point, help him or her in kindness and resources.

Minimum one should do is to give them publicity by distributing their books, videos and CD/DVDs.

At least make their Wikipedia presence colourful.

Chapter 20

Big Bang Conspiracy

The "Big Bang" was a big conspiracy to satisfy the church 100 years ago.

The church was very powerful and was dictatorial in its past.

Its first conspiracy was killing Socrates.

From then, there were many more episodes.

However, the most recent one was the hanging of the Vatican Banker under the London Bridge.

I was in London when that happened but police removed the body before I could have a look (with a forensic intent). The police never investigated it in full or published the details for public consumption.

Mafia had connection with Vatican and the death of one of the Popes soon after coming to office was not a mystery.

If one wants the details please read the "God Father Series" or see the films.

Before coming back to Big Bang conspiracy, let me summarize the current scientific knowledge, thanks to the Hubble Telescope (named after, one of the greatest scientists of yesteryear, who predicted the expansion of the universe).

Dark matter is roughly 25%.

Dark force 70%.

The matter we can see is only 5% (probably the God only saw this matter 7250 odd years ago-according to the Bible the life of this planet is little over 7000 years-and the Bible was written 2300 odd years ago).

Even in this 5% of matter, the byrons are not accounted for and are probably mingled among the gas clouds in the universe.

I have my own hypothesis and there are two books at Amazon if one is interested.

My hypothesis (not proven) is that the matter tends to settle to (transformation) a low energy level of dark matter (entropy) and immense amount of dark force.

If one push this theory far, all matter will disappear with time.

Byrons may be a manifestation.

But my counter theory proposes that dark matter and dark force turns back into matter right in the middle of the galaxies forming stars.

Why I am making these outrageous hypotheses is/are for the reader to look at fresh, the current theory including the "Big Bang".

We have not fully understood, the mechanics and a new form of physics (current physics, only applies to matter real) is necessary to understand the virtual matter (Dark Matter).

The Big Bang was proposed to satisfy the Church.

It is not the final theory.

There are two big holes in Big Bang theory.

I have explained them elsewhere.

The original matter comes from nothing.

Something cannot come from nothing and this supposition pleased the Church 100 years ago.

Then the time zero (time is a concept) was brought in to fit with the "something from nothing" and Big Bang was born and the god had an upper hand 100 years ago.

Not any more.

Both Albert Einstein and Fred Hoyle were under tremendous pressure from the church to suppress free thinking and inquiry.

That is why I say the church had been always dictatorial and helped dictatorial regimes in the past.

True scientist had no place and few were beheaded at the behest of the church.

When a new concept came in and if it did not fit in with the god creation, the church was at loggerhead with the emerging science.

If you read the Bible, god creation was added few centuries after its original scripture. So with new science coming into prominence the Bible has to be rewritten just like the Kuran does for many centuries.

There are two or three things that the Christianity has a monopoly.

1. Even Mafiosi boss can be excused by Redemption.

The god takes the crimes to his hand.

Penance is repentance of sins and Confession / Reconciliation. I believe one has to own his or her good or bad deeds in this world.

Prison is notorious option for criminals but there is no virtual heaven for good guys like us, just to see how it feels like to be in heaven.

Only option left for the good guys / girls in this world is some wine, I believe.

Religious parole is not in my book.

2. There are miracles.

3. Eternal hell (where is penance?) or heaven are undemocratic institutions.

One cannot go from one to the other (mutually antagonistic).

Chapter 21

Alternative Theories to "Big Bang"

For Paul Steinhardt and Neil Turok, the "Big Bang" ended on a summer day in 1999 in Cambridge, England. Sitting together at a conference they had organized, called "A School on Connecting Fundamental Physics and Cosmology", the two physicists suddenly hit on the same idea.

There are few holes in the Bing Bang.

Something cannot come from nothing.

Additionally what was the reason for its initiation.

Additionally how come time zero.

Maybe scientists were not ready to tackle this mystery and propose alternative possibilities.

What made the Big Bang go bang.

The deepest questions of all is:

What came before the Big Bang?

Steinhardt and Turok - working closely with a few like minded colleagues - have developed new insights into the problem and propose alternative to the prevailing, Genesis like view of cosmology.

According to the Big Bang theory, the whole universe emerged during a single moment some 13.7 billion years ago.

In the competing theory, our universe generates and regenerates itself an endless cycle of existence or becoming. The latest version of the cyclic model even matches key pieces of observational evidence supporting the older view.

This is the most detailed challenge yet to the old orthodoxy of the Big Bang.

Some researchers go further and envision a type of infinite time that plays out not just in this universe but in a multiverse - a multitude of universes, each with its own laws of physics and its own life story.

Still others seek to revise the very idea of time, rendering the concept of a "beginning" meaningless.

All of these cosmology heretics agree on one thing.

The Big Bang no longer defines the limit of how far the human mind can explore.

The Incredible Bulk

The latest elaboration of Steinhardt and Turok's cyclic cosmology, was published, spearheaded by Evgeny Buchbinder of Perimeter Institute for Theoretical Physics in Waterloo, Ontario. Yet the impulse behind this work far predates modern theories of the universe. In the fourth century A.D., St. Augustine pondered what the Lord was doing before the first day of Genesis (wryly repeating the exasperated retort that "He was preparing Hell for those who pry too deep").

The question became a scientific one in 1929, when Edwin Hubble determined that the universe was expanding. Extrapolated backward, Hubble's observation suggested the cosmos was flying apart from an explosive origin, the fabled Big Bang.

In the standard interpretation of the Big Bang, which took shape in the 1960s, the formative event was not an explosion that occurred at some point in space and time - it was an explosion of space and time.

In this view, time did not exist beforehand.

Even for many researchers in the field, this was a bitter pill to swallow.

It is hard to imagine how time just starting.

How does a universe decide when it is time to pop into existence?

For years, every attempt to understand what happened in that formative moment quickly hit a dead end. In the standard Big Bang model, the universe began in a state of near infinite density and temperature. At such extremes the known laws of physics break down. To push all the way back to the beginning of time, physicists needed a new theory, one that blended general relativity with quantum mechanics.

The prospects for making sense of the Big Bang began to improve in the 1990s as physicists refined their ideas in string theory, a promising approach for reconciling the relativity and quantum views.

Nobody knows yet whether string theory matches up with the real world - the Large Hadron Collider, a particle smasher, may provide some clues. It has already formulated, stunning new ideas how the universe is constructed. Most notably, current versions of string theory posit seven hidden dimensions of space in addition to the three we experience. Strange and wonderful things can happen in those extra dimensions.

That is what inspired Steinhardt (of Princeton University) and Turok (of Cambridge University) to set up their fateful conference in 1999.

"We organized the conference because we both felt that the standard Big Bang model was failing to explain things.

Turok says.

"We wanted to bring people together to talk about what string theory could do for cosmology."

The key concept turned out to be a "brane," a three-dimensional world embedded in a higher dimensional space (the term, in the language of string theory, is just short for membrane). "People had just started talking about branes when we set up the conference," Steinhardt recalls.

Together Neil and I went to a talk where the speaker was describing them as static objects. Afterward we both asked the same question:

What happens if the branes can move?

What happens if they collide?"

A remarkable picture began to take shape in the two physicists' minds. A sheet of paper blowing in the wind is a kind of two-dimensional membrane tumbling through our three-dimensional world. For Steinhardt and Turok, our entire universe is just one sheet, or 3-D brane, moving through a four dimensional background called "the bulk."

Our brane is not the only one; there are others moving through the bulk as well. Just as two sheets of paper could be blown together in a storm, different 3-D branes could collide within the bulk.

The equations of string theory indicated that each 3-D brane would exert powerful forces on others nearby in the bulk. Vast quantities of energy lie bound up in those forces.

A collision between two branes could unleash those energies.

From the inside, the result would look like a tremendous explosion. Even more intriguing, the theoretical characteristics of that explosion closely matched the observed properties of the Big Bang - including the cosmic microwave background, the afterglow of the universe's fiercely hot early days.

"That was amazing for us because it meant colliding branes could explain one of the key pieces of evidence people use to support the Big Bang," Steinhardt says.

Three years later came a second epiphany:

Steinhardt and Turok found their story did not end after the collision.

"We weren't looking for cycles," Steinhardt says, "but the model naturally produces them."

After a collision, energy gives rise to matter in the brane worlds. The matter then evolves into the kind of universe we know: galaxies, stars, planets, the works. Space within the branes expands, and at first the distance between the branes (in the bulk) grows too.

When the brane worlds expand so much that their space is nearly empty, however, attractive forces between the branes draw the world sheets together again.

A new collision occurs, and a new cycle of creation begins.

In this model, each round of existence - each cycle from one collision to the next - stretches about a trillion years. By that reckoning, our universe is still in its infancy, being only 0.1 percent of the way through the current cycle.

The cyclic universe directly solves the problem of before.

With an infinity of Big Bangs, time stretches into forever in both directions.

"The Big Bang was not the beginning of space and time," Steinhardt says.

"There was a before, and before matters because it leaves an imprint on what happens in the next cycle."

Not everyone is pleased by this departure from the usual cosmological thinking.

Some researchers consider Steinhardt and Turok's ideas misguided or even dangerous.

"I had one well-respected scientist tells me we should stop because we were undermining public confidence in the Big Bang," Turok says.

But part of the appeal of the cyclic universe is that it is not just a beautiful idea - it is a testable one.

The standard model of the early universe predicts that space is full of gravitational waves, ripples in space-time left over from the first instants after the Big Bang.

These waves look very different in the cyclic model, and those differences could be measured - as soon as physicists develop an effective gravity-wave detector.

"It may take 20 years before we have the technology," Turok says, "but in principle it can be done.

Given the importance of the question, I'd say it's worth the wait."

Time's Arrow

While the concept of a cyclic universe provides a way to explore the Big Bang's past, some scientists believe that Steinhardt and Turok have skirted the deeper issue of origins.

"The real problem is not the beginning of time but the arrow of time," says Sean Carroll, a theoretical physicist at Caltech.

"Looking for a universe that repeats itself is exactly what you do not want. Cycles still give us a time that flows with a definite direction, and the direction of time is the very thing we need to explain."

In 2004 Carroll and a graduate student of his, Jennifer Chen, came up with a much different answer to the problem of before. In his view, time's arrow and time's beginning cannot be treated separately.

There is no way to address what came before the Big Bang until we understand why the before precedes the after. Like Steinhardt and Turok, Carroll thinks that finding the answer requires rethinking the full extent of the universe, but Carroll is not satisfied with adding more dimensions.

He also wants to add more universes - a whole lot more of them - to show that, in the big picture, time does not flow so much but advance symmetrically backward and forward.

The one-way progression of time, always into the future, is one of the greatest enigmas in physics. The equations governing individual objects do not care about time's direction.

Imagine a movie of two billiard balls colliding.

There is no way to say, if the movie is being run forward or backward. But if you gather a zillion atoms together in something like a balloon, past and future look very different. Pop the balloon and the air molecules inside quickly fill the entire space. They never race backward to reinflate the balloon.

In any such large group of objects, the system trends toward equilibrium. Physicists use the term entropy to describe how far a system is from equilibrium. The closer it is, the higher its entropy; full equilibrium is, by definition, the maximum value. So the path from low entropy (all the molecules in one corner of the room, unstable) to maximum entropy (the molecules evenly distributed in the room, stable) defines the arrow of time.

The route to equilibrium separates before from after.

Once you hit equilibrium the arrow of time no longer matters, because change is no longer possible.

"Our universe has been evolving for 13 billion years," Carroll says, "so it clearly did not start in equilibrium."

Rather, all the matter, energy, space, and even time in the universe must have started in a state of extraordinarily low entropy. That is the only way we could begin with a Big Bang and end up with the wonderfully diverse cosmos of today.

Understand how that happened, Carroll argues, and you will understand the bigger process that brought our universe into being.

To demonstrate just how strange our universe is, Carroll considers all the other ways it might have been constructed. Thinking about the range of possibilities, he wonders:

Why did the initial setup of the universe allow cosmic time to have a direction?

There are an infinite number of ways the initial universe could have been set up. An overwhelming majority of them have high entropy.

These high-entropy universes would be boring and inert; evolution and change would not be possible. Such a universe could not produce galaxies and stars, and it certainly could not support life.

It is almost as if our universe were fine-tuned to start out far from equilibrium so it could possess an arrow of time. But to a physicist, invoking fine-tuning is akin to saying "a miracle occurred."

For Carroll, the challenge was finding a process that would explain the universe's low entropy naturally, without any appeal to incredible coincidence or worse to a miracle.

Carroll found that process hidden inside one of the strangest and most exciting recent elaborations of the Big Bang theory. In 1984, MIT physicist Alan Guth suggested that the very young universe had gone through a brief period of runaway expansion, which he called "inflation," and that this expansion had blown up one small corner of an earlier universe into everything we see. In the late 1980s Guth and other physicists, most notably Andrei Linde, now at Stanford, saw that inflation might happen over and over in a process of "eternal inflation."

As a result, pocket universes much like our own might be popping out of the uninflated background all the time.

This multitude of universes was called, inevitably, the multiverse.

Carroll found in the multiverse concept a solution to both the direction and the origin of cosmic time. He had been musing over the arrow of time as far back as graduate school in the late 1980s, when he published papers on the feasibility of time travel using known physics. Eternal inflation suggested that it was not enough to think about time in our universe only; he realized he needed to consider it in a much bigger, multiverse context.

"We wondered if eternal inflation could work in both directions," Carroll says.

"That means there would be no need for a single Big Bang. Pocket universes would always sprout from the uninflated background. The trick needed to make eternal inflation work was to find a generic starting point. An easy-to-achieve condition that would occur infinitely many times and allow eternal inflation to flow in both directions."

A full theory of eternal inflation came together in Carroll's mind in 2004, while he was attending a five month workshop on cosmology at the University of California at Santa Barbara's famous Kavli Institute of Theoretical Physics with his student Jennifer Chen.

"You go to a place like Kavli and you are away from the normal responsibilities of teaching," Carroll says.

"That gives you time to pull things together."

In those few months, Carroll and Chen worked out a vision of a profligate multiverse without beginnings, endings, or an arrow of time.

"All you need," Carroll says, with a physicist's penchant for understatement, "Is to start with some empty space, a shard of dark energy, and some patience."

Dark energy - a hidden type of energy embedded in empty space, whose existence is strongly confirmed by recent observations - is crucial because quantum physics says that any energy field will always yield random fluctuations. In Carroll and Chen's theory, fluctuations in the dark energy background function as seeds that trigger new rounds of inflation, creating a crop of pocket universes from empty space.

"Some of these pocket universes will collapse into black holes and evaporate, taking themselves out of the picture" Carroll says.

"But others will expand forever. The ones that expand eventually thin out. They become the new empty space from which more inflation can start."

The whole process can happen again and again. Amazingly, the direction of time does not matter in the process.

"That is the funny part. You can evolve the little inflating universes in either direction away from your generic starting point," Carroll says. In the super-far past of our universe, long before the Big Bang, there could have been other Big Bangs for which the arrow of time ran in the opposite direction.

On the grandest scale, the multiverse is in the form of interconnected pocket universes, completely symmetric with respect to time. Some universes move forward, but overall, an equal number move backward. With infinite space in infinite universes, there are no bounds on entropy.

It can always increase; every universe is born with room (and entropy) to evolve. The Big Bang is just our Big Bang, and it is not unique. The question of before melts away because the multiverse has always existed and always will, evolving but - in a statistical sense - always the same.

After completing his multiverse paper with Chen, Carroll felt a twinge of dismay.

"When you finish something like this, it's bittersweet.

The fun with hard problems can be in the chase," he says. Luckily for him, the chase goes on.

"Our paper really expresses a minority viewpoint," he admits. He is now hard at work on follow up papers fleshing out the details and bolstering his argument.

The Nows Have It

In 1999, while Steinhardt and Turok were convening in Cambridge and Carroll was meditating on the meaning of the multiverse, rebel physicist Julian Barbour published The End of Time - a manifesto suggesting that attempts to address what came before the Big Bang were based on a fundamental mistake.

There is no need to find a solution to time's beginning, Barbour insisted, because time does not actually exist.

Back in 1963, a magazine article had changed Barbour's life. At the time he was just a young physics graduate student heading off for a relaxing trip to the mountains.

"I was studying in Germany and had brought an article with me on holiday to the Bavarian Alps," says Barbour, now 71. "It was about the great physicist Paul Dirac. He was speculating on the nature of time and space in the theory of relativity."

After finishing the article Barbour was left with a question he would never be able to relinquish:

What, really, is time?

He could not stop thinking about it. He turned around halfway up the mountain and never made it to the top.

"I knew that it would take years to understand my question," Barbour recalls.

"There was no way I could have a normal academic career, publishing paper after paper, and really get anywhere."

With bulldog determination he left academic physics and settled in rural England, supporting his family translating Russian scientific journals. Thirty eight years later, still living in the same house, he has worked out enough answers to rise from obscurity and capture the attention of the world's physics community.

In the 1970s Barbour began publishing his ideas in respected but slightly unconventional journals, like The British Journal for the Philosophy of Science and Proceedings of the Royal Society A. He continues to issue papers, most recently with his collaborator Edward Anderson of the University of Cambridge. Barbour's arguments are complex, but his core idea remains simplicity itself:

There is no time.

"If you try to get your hands on time, it's always slipping through your fingers," Barbour says with his disarming English charm.

"My feeling is that people can't get hold of time because it isn't there at all."

Isaac Newton thought of time as a river flowing at the same rate everywhere.

Albert Einstein unified space and time into a single entity, but he still held on to the concept of time as a measure of change.

In Barbour's view there is no invisible river of time.

Instead, he thinks that change merely creates an illusion of time, with each individual moment existing in its own right, complete and whole.

He calls these moments "Nows."

"As we live, we seem to move through a succession of Nows.

The question is, what are they?" Barbour asks.

His answer.

Each Now is an arrangement of everything in the universe.

"We have the strong impression that things have definite positions relative to each other. I aim to abstract away everything we cannot see, directly or indirectly, and simply keep this idea of many different things coexisting at once. There are simply the Nows, nothing more and nothing less."

Barbour's Nows can be imagined as pages of a novel ripped from the book's spine and tossed randomly onto the floor. Each page is a separate entity. Arranging the pages in some special order and moving through them step by step makes it seem that a story is unfolding. Even so, no matter how we arrange the sheets, each page is complete and independent. For Barbour, reality is just the physics of these Nows taken together as a whole.

"What really intrigues me is that the totality of all possible Nows has a very special structure," he says. "You can think of it as a landscape or country. Each point in this country is a Now, and I call the country Platonia," in reference to Plato's conception of a deeper reality, "because it is timeless and created by perfect mathematical rules.

Platonia is the true arena of the universe."

In Platonia all possible configurations of the universe, every possible location of every atom, exist simultaneously. There is no past moment that flows into a future moment; the question of what came before the Big Bang never arises because Barbour's cosmology has no time.

The Big Bang is not an event in the distant past; it is just one special place in Platonia. Our illusion of the past comes because each Now in Platonia contains objects that appear as "records," in Barbour's language.

"The only evidence you have of last week is your memory - but memory comes from a stable structure of neurons in your brain now. The only evidence we have of the earth's past are rocks and fossils - but these are just stable structures in the form of an arrangement of minerals we examine in the present. All we have are these records, and we only have them in this Now," Barbour says.

In his theory, some Nows are linked to others in Platonia's landscape even though they all exist simultaneously. Those links create the appearance of a sequence from past to future, but there is no actual flow of time from one Now to another.

"Think of the integers," Barbour says.

"Every integer exists simultaneously.

But some of the integers are linked in structure, like the set of all primes or the numbers you get from the Fibonacci series."

Yet the number 3 does not occur in the past of the number 5 any more than the Big Bang exists in the past of the year 2008.

These ideas might sound like the stuff of late-night dorm-room conversations, but Barbour has spent four decades hammering them out in the hard language of mathematical physics.

He has blended Platonia with the equations of quantum mechanics to devise a mathematical description of a "changeless" physics. With Irish collaborator Niall Ó Murchadha of the National University of Ireland in Cork, Barbour is continuing to reformulate a time-free version of Einstein's theory.

So What Really Happened?

For each of the alternatives to the Big Bang, it is easier to demonstrate the appeal of the idea than to prove that it is correct. Steinhardt and Turok's cyclic cosmology can account for critical pieces of evidence usually cited to support the Big Bang, but the experiments that could put it over the top are decades away.

Carroll's model of the multiverse depends on a speculative interpretation of inflationary cosmology, which is itself only loosely verified.

Barbour stands at the farthest extreme. He has no way to test his concept of Platonia. The power of his ideas rests heavily on the beauty of their formulation and on their capacity to unify physics. "What we are working out now is simple and coherent," Barbour says, "and because of that I believe it is showing us something fundamental."

The payoff that Barbour offers is not just a mathematical solution but a philosophical one. In place of all the conflicting notions about the Big Bang and what came before, he offers a way out. He proposes letting go of the past - of the whole idea of the past - and living fully, happily, in the Now. In one model, each round of existence stretches a trillion years. By that reckoning, our universe is still in its infancy.

Chapter 22

Global Warming

America aided by Oil Companies headed the misinformation campaign. Everybody knows global warming is due to fossil fuels and they dragged their feet until the last moment. Even the protocol signed recently has major loop holes for America, Brazil and China to slip through.

Instead of boring the reader with technical aspects, I would pen down my own thinking and personal experiences below especially related to the pet fish and my water plants.

Population Expansion and Global Warming

The global warming has direct and indirect relationship to human population expansion, not necessarily due to increased consumption of fossil fuel alone but due to other factors. With the population expansion the demand for resources increases exponentially. The felling of trees is one example for instance. Food that we consume and the demand for food may have relationship with global warming. The changing agricultural practices to overcome supply and demand problem with monoculture and cash crops, not only affect the biodiversity, it probably contributes to global warming. Meat consumption and possibly the increased consumption of chicken have relationship with global warming. It cannot be single out but multitude of factors contribute to global warming.

I believe we have come to the optimum population of 6 billions.

The earth cannot sustain 9 billions.

This extra 3 billion will upset the balance of biophysical relationships that come into equilibrium in the microscopic levels or the nano-level.

The mega level is man and his megalomaniac beahviour.

Earth cannot sustain its biophysical relationships with the rate at which the human population expands and consumes.

Biophysical barrier will break down soon.

Then calamities after calamities would occur.

Gloom, bloom and doom, unless we arrest the population expansion and retard the rate of use of easily available energy resources.

First, we have to arrest the population growth.

The second, clean water for both man and beings including plants.

Third, we have to have the food security.

Fourth, we have to prevent the made made causes of global warming. Even in this country the priority is on the energy and its use, its expansion and not conservation.

Fifth, we should invest on alternative power sources.

We got our priorities wrong.

We will loose all our biodiversity.

The rate at which we domesticate elephants there won't be any left in the wild.

Forest Harvesting

I was bit inquisitive why there are so many tornadoes and hurricanes in America.

I just went to Google Earth and had a little peep from above of North America, the West Coast and the East Coast.

There is hardly any difference in tree cover over the land, East or West. Mostly farmland and build up areas.

That did not give me any clue to the state of the forest cover. Then I went and searched deforestation.

Americans harvested 90% of the land in 70 years from 1850 to 1920. Entire East was covered with forest and fair proportion of the West was covered with primary forest. The deforestation continues to this century and America now has mainly secondary forest covering only 10% of the land.

Americans knew that the CO_2 problem started around 1920s and continues even to present day due to their over exploitation of fossil fuel.

Did they tell the truth to the world?

Big No.

In Sri-Lanka we had 90% forest cover until around 1850 and British started deforestation for coffee and tea cultivation. By the time they left in 1948 forest cover was over 60% but before they left they passed a law prohibiting encroachment of the Crown Land.

From 1948 to 2000 we have decimated another 40 percent especially after 1970. We are now below the minimal threshold of 25% to maintain our rivers.

This land now can be called the People's Land instead of the Crown Land and the tree felling and the development go on. With the thermal (coal) power plants in full operation, in less than two decades, we will be approaching 10% level which is the cut off point for desert classification.

Acid rain will do the rest even if we stop cutting trees to Zero.

Then we can say we are better than America in case of deforestation and go for an IMF loan.

Instead of wasting time, with boring exposition, of true facts, let me expand my own personal experience which I recorded many moons ago.

My Exotic Tropical Fish

I was expecting a catastrophe.

It surely did come.

One day of negligence all but four of my Neon Tetras were gone. My final diagnosis global warming (perhaps enhanced by our coal power production).

Apparently, England has had the warmest springs.

Summer heat is yet to come.

Three decade ago, UK was closing all the mines and decommissioning all the Coal Power units, mainly due to health issues. In our country, the health minister is obsessed with cigarettes but does not utter a word about coal power and its emissions. One hour of coal power is probably equal to ten to hundred times the tobacco smoked by all Sri-Lankans.

That is my rough formula.

He has got his priorities mixed up, I believe.

I have talked about coal power, pollution and global warming for decades but when it affected my pet fish, even a few lines of reference is desired.

Tropical fish survive between 76 to 82 Fahrenheit (24 to 30 degrees Centigrade).

The temperature in my roof top garden is over 95 Fahrenheit (32 degrees) and above and my guppies were dying, in numbers. in spite of the fact, that they were in the shade.

Mind you this is Peradeniya only a walking distance from Botanical garden.

As a precaution green swordtail (hard to find now) were taken to a shady tank. I had a big problem of increasing their numbers from a tiny stock (five years of trouble) and they do breed now, not as profusely, as guppy fish.

My guppies were down to 5 in number, due to global warming 10 years ago.

That was an epic episode.

Now I have a beautiful collection of them outdoors, mainly to control mosquitoes. The top of all my outdoor fish containers are either covered with floating plants or black polythene nets to reduce the penetration of sunlight.

My real interest is on water plants with beautiful tiny flowers which bloom almost every day due to warm weather.

The guppy fish started dying in numbers again.

I was not happy and I decided to keep a tiny stock indoors, in case the previous epic episode repeat itself and they will be down to five again.

I decided to move the fish indoors to a new plastic tank.

From day one, there were problems.

I had some exotic water plants and the algae were invading the tank. They even discolored the leaves of the water plants.

The guppy started breeding in their numbers in the new and safe water front. The Neon tetras were outnumbered.

All remained in good health.

But one day of lapse on my part and with water quality going down, most of the neon tetras were gone and some were making their last breaths.

I did all what I could, quickly changed the water but only four survived.

The casualty included one algae eater.

Few guppies were included.

I was out and came late and did not bother to have a casual look at the fish tank. They would have been showing signs of ill health which I missed.

In good old days, one could leave the well maintained fish tank for six to eight weeks without changing water. Now I have to change water every 7 to 10 days. If I miss or forget to change the water by a single day most of them will perish.

Now I regret myself for having two indoor fish tanks.

Looking after them is a big hassle!

If a major catastrophe intervenes, I am going to stop my hobby for good.

Not only tropical fish even the tropical plants are dying.

There is no chance for my water plants to survive, if I take a week holiday.

It is all due to global warming.

It is mostly a man made catastrophe.

Chapter 23

Mad Cow Disease

The origin of this disease and from which country it originated is at best a conjecture.

First, the disease was neither identified nor diagnosed in the cattle or in humans.

Second, is that the scientist conspired not to reveal it until it almost became an endemic disease in United Kingdom.

Third, meat industry was an important export industry in 1980s.

Fourth, the politicians connived with the industry and suppressed the scientific know how for a prolonged period, since an infectious agent was not discovered.

Fifth, unorthodox practice of mixing brain and offal with pasture to increase meat production was found to be course.

Sixth, it is believed (their brain and offal were mixed with pasture and fed to cattle) that it is a slow virus that infected the brain of sheep.

Seventh, the the formation of prion protein by mis-folding of normal protein does occur after ingestion of partially digested bovine protein. This bovine protein act as a template for mis-folding of protein in the brain. It acts as a non viral infectious agent.

Eighth it has a long incubation period.

Ninth, it causes dementia and there is no treatment.

How long have health officials been concerned about Mad Cow Disease?

Mad Cow Disease has been of great concern since 1986, when it was first reported among cattle in the U.K. At its peak in January 1993, almost 1,000 new cases per week were identified. Concern about this disease grew significantly in 1996 when an association between mad cow disease and vCJD in humans was discovered.

What other countries have reported cases of Mad Cow Disease?

The disease also has been confirmed in cattle born in Austria, Belgium, Czech Republic, Denmark, Finland, France, Germany, Italy, Ireland, Israel, Japan, Liechtenstein, Luxembourg, Netherland, Poland, Portugal, Slovakia, Slovenia, Spain, Switzerland, and United Kingdom.

Canada has also been added to the list of countries from which imports are restricted to USA, although that ban has been lifted recently. Importation of minimal risk meat products is now allowed from Canada.

Does Mad Cow Disease affect humans?

A human version of mad cow disease called variant Creutzfeldt-Jakob disease (vCJD) is believed to be caused by eating beef products contaminated with central nervous system tissue, such as brain and spinal cord, from cattle infected with mad cow disease. For this reason, the USDA requires that all brain and spinal cord materials be removed from high risk cattle, older cattle, animals that are unable to walk, and any animal that shows any signs of a neurological problem.

These cow products do not enter the U.S. food supply. The USDA believes this practice effectively safeguards U.S. public health from vCJD.

Bovine spongiform encephalopathy (BSE), commonly known as Mad Cow Disease, is a fatal neurodegenerative disease (encephalopathy) in cattle that causes a spongy degeneration of the brain and spinal cord. BSE has a long incubation period, about 2.5 to 8 years, usually affecting adult cattle at a peak age onset of four to five years, all breeds being equally susceptible.

BSE is caused by a mis-folded protein—a prion.

In the United Kingdom, the country worst affected by an epidemic in 1986-98, more than 180,000 cattle were infected and 4.4 million slaughtered during the eradication program.

The disease may be most easily transmitted to human beings by eating food contaminated with the brain, spinal cord or digestive tract of infected carcasses. However, the infectious agent, although most highly concentrated in nervous tissue, can be found in virtually all tissues throughout the body, including blood.

In humans, it is known as new variant Creutzfeldt–Jakob disease (vCJD or nvCJD), and by June 2014 it had killed 177 people in the United Kingdom, and 52 elsewhere.

Between 460,000 and 482,000 BSE-infected animals had entered the human food chain before controls on high risk offal were introduced in 1989.

A British and Irish inquiry into BSE concluded the epizootic was caused by cattle, which are normally herbivores, being fed the remains of other cattle in the form of meat and bone meal (MBM), which caused the infectious agent to spread.

The cause of BSE may be from the contamination of MBM from sheep with scrapie that were processed in the same slaughterhouse. The epidemic was probably accelerated by the recycling of infected bovine tissues prior to the recognition of BSE.

The origin of the disease itself remains unknown.

The infectious agent is distinctive for the high temperatures at which it remains viable, over 600 °C (about 1100 °F). This contributed to the spread of the disease in the United Kingdom, which had reduced the temperatures used during its rendering process.

Another contributory factor was the feeding of infected protein supplements to very young calves.

Chapter 24

Coconut Theory

America's peanut industry killed our coconut industry for over 50 years. The American President Jimmy Carter was a peanut farmer.

The conspiracy propaganda, is to say all bad things about coconut oil.

It has cholesterol and cholesterol is bad for health.

It has saturated fats.

Do spurious testes on hydrogenated coconut oil (not natural oil) and extrapolate the results to natural coconut oil.

Industrial hydrogenation of coconut oil to margarine is the main conspiracy.

Promote and perpetuate unsaturated fatty acid is good for your heart. It is a fallacy.

Promote palm oil industry instead of coconut oil.

It has PCBs.

Because the conspiracies were propagated over 50 years and most of the educated souls have come to senses with reality and aware of the current and past trends, no point repeating old phrases and misconceptions and perpetuating further, the grand conspiracy, but let me bare bones and tell you the health benefits of coconut oil.

Coconut Experience

It is now believed that coconut oil is the healthiest of oils on earth.

Some even say, it is the miracle oil, if only prepared as virgin coconut oil free of harmful contaminants.

It is the virgin coconut oil that has this unique property.

My intention is to revive the age old tradition of extracting coconut oil for domestic consumption.

Virgin coconut oil is an extract of pure coconut oil free of polychlorinated biophenols. All industrial coconut oils except virgin coconut oil are contaminated with certain amount of these chemicals during industrial processing.

PCBs

PCBs are known to increase the incidence and risk of breast cancer.

PCBs, or polychlorinated biphenyls, are a class of chemicals with a variety of industrial and commercial applications. Because of their health concerns these chemicals (PCBs) are banned in the US and Canada but they still linger in the environment and are present in the food chain, particularly in fatty foods. The major culprit of the production of this chemical is the chlorination of water before the so called purification (only from bacteria). Our doctors forget about the chemical contaminants in our water because of their allopathic inclination to produce and distribute in mass scale, chemicals of dubious nature in the name of treatment of modern diseases.

PCBPs

I have renamed these items (PCBPs) in a generic descriptive term of polluted when processing (P), chemically contaminated (C) while preserving and biologically bombarded (degrading) for storage and mass scale production (P) by commercial chains and companies (including hydrogenation of natural oil for long storage life).

Emerging research on the nutritional and medicinal benefits on coconut oil has surfaced in recent years. Much of that research has been done by Dr. Mary Enig who has classified coconuts as a "functional food," which provides health benefits over and beyond the basic nutrients. She has specifically identified lauric acid as a key ingredient in coconut products.

Approximately 50% of the fatty acids in coconut fat are lauric acid. Lauric acid is a medium chain fatty acid, which has the additional beneficial function of being formed into monolaurin in the human or animal body. Monolaurin is the antiviral, antibacterial, and anti-protozoal monoglyceride used by the human or animal to destroy lipid coated viruses such as HIV, herpes, cytomegalovirus, influenza, various pathogenic bacteria including listeria monocytogenes and helicobacter pylori, and protozoa such as giardia lamblia.

Some studies have also shown some antimicrobial effects of the free lauric acid."

Coconut oil has been used for centuries as a vital source of food for health and general wellbeing in traditional communities in the tropics including Sri-Lanka.

Recent research verifies traditional wisdom and beliefs that the coconut palm is "The Tree of Life" and that, just like any other pure, whole food, coconuts and coconut virgin oils have a significant role to play in a well balanced, nutritious diet.

Abandoning unhealthy lifestyles and reverting to natural foods can help to reverse many of the diseases that manifest through the unhealthy living styles in the modern society and the consumption of highly refined foods in our diet.

Our daily food intake should contain various combinations of fats or oils. However, the structures of different oils are as diverse as nature itself.

A basic knowledge of how fats are classified by their chemical nature is necessary to understand the logic of recommended foods for consumption but I am not interested in teaching biochemistry but to educate the masses how to prepare virgin coconut oil and consume a healthy meal.However, this classification is important when choosing oil to augment and support a healthy lifestyle for our children and families.

Research shows that replacing other cooking oils with virgin coconut oil generally creates a more favorable HDL/LDL ratio.

This oil has antiviral, antibacterial, antimicrobial, and anti-protozoal properties and, like all whole foods, contains nutrients for a healthy body.

A Case for Coconut oil

Coconut oil is in many ways a unique gift of nature.

It contains 62% Medium Chain Fatty Acids, even though they are saturated fatty acids, they are not harmful to our heart. MCFAs give coconut oil an important property which other saturated and unsaturated fats lack.

It is often lost under the dusty political cloud of the cholesterol debate, which is an American conspiracy.

Quasi immune to light oxidation and highly resistant to rancidity the oil is functional as a safe nutritional source in most climates without the need for refrigeration or special storage conditions.

However, the fact that is lost in the furor over cholesterol is the differentiation of the different sub-groups of fatty acids and in particular the importance the of Medium Chain Fatty Acids. Modern researchers have discovered important health benefits and critical bodily functions that are supported by the pool of MCFAs available to the human body.

The high concentration of Medium Chain Fatty Acids (MCFAs) of 62% is the critical aspect of why this tropical coconut oil not only behaves differently to any other saturated fat, but is also deemed healthier than most unsaturated oils, the latter often having a much higher concentration of long chain fatty acids.

Generally MCFAs are of a smaller particle size thus requiring less energy and fewer enzymes for absorption, placing less strain on the digestive system.

This is further enhanced by the fact that, short and medium chain fatty acids are made soluble and absorbed in the aqueous phase of the intestinal contents and transported directly via the portal vein to the liver where they are consumed as an energy source similar to carbohydrates. MCFAs have also been shown to assist the absorption and retention of calcium, magnesium and some amino acids as well supporting the healthy functioning of the thyroid.

Notwithstanding, all the bodily functions that may be enhanced by the consumption of MCFAs the most outstanding finding of modern research has been the capacity of certain MCFA to enhance the human immune system and actively participate in the capacity of the human body to fight virus, bacteria and fungi.

The most important of these MCFAs that enhance and boost the body's immune system are Caproic Acid, Caprylic Acid, Capric Acid and Lauric Acid and Myristic Acid.

It is however the powerful capacity of Lauric acid to immobilize harmful invaders of our body, which has excited researchers the most.

Although Dr Jon Kabara had already noted the antimicrobial activity of lauric acid in 1966, it has been only in the last 20 years that the potential of all MCFA in fighting bacterial and viral infections has been realized.

Furthermore, in addition to their capacity to inactivate pathogenic organisms, the use of MCFAs has shown no negative toxicological or pharmacological side effects.

If anything MCFA have shown to weaken viruses or bacteria that show antibiotic resistance to such an extent that these drugs can actually function against these organisms, while the use of MCFAs over time has not shown any build-up of resistance by microbial organisms to these fatty acids.

In practical terms, the consumption of MCFAs would not only boost the human immune system without fear of negative side effects, but also that they could be consumed to compliment the functions of general antibiotic treatments when their effectiveness is decreased by built up resistance found in bacteria. MCFAs are effective mostly in their digested form or in the case of lauric acid: monolaurin.

The effectiveness of monolaurin is based on the small size of the particle and the similarity in substance to the lipid (fat) membranes coating harmful microorganisms. Monolaurin is attracted to the virus and is then easily absorbed into its quasi fluid membrane weakening it to such a degree that it literally splits open, killing the organism and allowing the bodies own white blood cells to effectively clean and dispose of the remnants.

These research results have lead to the incorporation of especially Caprylic, Capric and Lauric Acid into many accepted anti-viral/anti-bacterial pharmaceutical treatments today.

The fact is that over 62% of coconut oil is made up of these three major fatty acids.

In addition the majority, around 48%, is lauric acid, the fundamental building block of our bodies immune system and the most effective anti-pathogenic of all MCFAs.

Coconut Saga

Coconut Oil Fuels Your Metabolism!

Coconut Oil Boosts Your Thyroid!

Coconut Oil Protects and Beautifies Your Skin!

Coconut Oil Can Save Your Brain!

Coconut Oil is Amazing for Heart Health!

One must replace "Coconut Oil" with "Pol Kiri" and "Pol Sambol".

However, Oil can be used on hair and skin liberally.

Coconut Oil Fuels Your Metabolism!

Coconut oil consist of Medium Chain Fatty Acids (MCFAs). It is a ready made energy source which can replace sugar (10 grams sugar = 1 gram of Coconut Oil roughly) weight for weight.

One does not become hungry.

It does not stimulate insulin.

It gives more energy and it is not stored as body fat.

Body fat is long chain fatty acids.

It is useful both in Diabetes and Hypothyroidism.

In diabetes it replaces carbohydrates as an energy source, so Insulin requirement is reduced. In hypothyroidism since it provides freely available energy, it cuts down the need for more thyroxine for metabolism to keep warm. Not only that it probably helps in hormonal balance (homeostasis) of the thyroid. Goiter is rare in countries where coconut oil is the staple oil. In pre-diabetes and early hypothyroidism it retards obesity due to above two actions.

I am not sure about its role in normal adults.

There is a caveat, if you take more "Oil" than YOU need for your metabolism, all the benefits are nullified.

There are published research work that support above statements. According to studies, the effects of coconut oil on abdominal fat is amazing!

A study published in 2009, in the Journal of Lipids consisted of testing the effects of either 2 tablespoons of coconut oil or 2 tablespoons of soybean on a group of 40 women over the span of 28 days. Results showed that the group that ate the coconut oil had a decrease in abdominal fat, while the soybean oil group actually showed a slight increase in belly fat. Additionally, the group that ate the coconut oil showed increased HDL 'good' cholesterol levels, while the soybean oil group had decreased HDL cholesterol and increased LDL ' bad' cholesterol.

The Journal of Nutrition published a study where researchers investigated all studies relative to medium chain fatty acids (MCFAs) that are abundant in coconut fat and weight management.

The studies showed that diets rich in fats such as those found in coconut oil prompted a boost in metabolism, increase in energy, decrease in food consumption, reduced body weight and lower body fat mass.

The study authors highly recommend using oils that contain MCFAs, such as coconut oil, as a tool to drop extra abdominal fat, manage a healthy weight, and even as a way to treat obesity.

Yet another study that assessed body belly fat weight and fat storage relative to three different types of diets including a low-fat diet, high-fat diet with long chain fatty acids (LCFAs) and a high fat diet with MCFAs.

In order to bring about weight gain, caloric intakes were adjusted for the diets.

At the end of the research period (which lasted 44 days), the low-fat diet group stored an average of 0.47 grams of fat per day, the LCFA group stored 0.48 grams of fat per day, and the MCFA group only stored a mere 0.19 grams per day (despite purposely increasing calories).

Those in the MCFA group (coconut fat) had a 60 percent reduction in body fat stored compared to the other diets. Another added bonus of consuming coconut oil (and coconut cream or milk) is that it tends to make us feel fuller for longer. Studies indicate that MCFAs help increase feelings of fullness and lead to a reduction in calorie intake when compared to the same amount of calories from other fats. When MCFAs are metabolized, ketone bodies are created in the liver – these have been shown to have a strong appetite reducing effect helping you to lose fat faster.

One should avoid other fats including hydrogenated coconut oil.

For thousands of years, coconuts have been a staple of tropical cuisines, and those who followed a traditional coconut-based diet, such as Pacific Islanders, had none of the heart disease, cancer, diabetes, or other illnesses that plague modern America (at least until they switched to Western oils and diet).

Coconut had a solid reputation in America in the first half of the last century, at least for those who knew about, especially after it's benefits became well known from soldiers in the Pacific Theater of World War II.

However, this all changed in 1950s, from "Three "Tragic" Research Papers", whose data were doctored to suit American Conspiracy.

One lady researcher in America fed mice with hydrogenated (not pure coconut oil) coconut oil and doctored a paper to say coconut oil is bad to promote Peanut Oil. She committed suicide when it was widely known that she manipulated the data.

In 1954, another researcher (D. K.) published two academic papers. The initial research described the effects of feeding cholesterol to rabbits and indicated that this may lead to the formation of blocked arteries and thus contribute to potential heart disease.

In his second paper he described the beneficial effects of consuming polyunsaturated fatty acids from the oil of corn, soybeans, safflower and sunflower seeds for the lowering, at least temporarily, of cholesterol in the blood.

Although many studies at a later time had also shown research to the contrary, the myth stuck and by the mid 60's the reputation for saturated oils in America all but destroyed.

Then came the "Marketing Machine" of big business.

The well oiled marketing machinery funded by the soy bean and corn industry and supported by the American Heart

Association was committed to change the American Diet, calling among others, for the substitution of saturated fats for polyunsaturates. Today worldwide heart disease is still on the increase and obesity, linked to the American diet is a major social problem.

All saturated fats are not created equal.

The travesty of these actions was that one of nature's most amazing resources, tropical oils, and especially coconut oil with all its functional, nutritional and pharmaceutical possibilities, has been lost to modern medicine for decades.

Although saturated, coconut oil is structurally, pharmaceutically and behaviorally different to any other natural oil or fat. Coconut Oil is one of the healthiest things you can put in (and on) your body!

What are the health benefits of coconut oil being reported?

Some of the most recent research has come from people suffering from senile dementia (Alzheimer's Disease), with reports of people improving or even reversing the effects of Alzheimer's by using coconut oil, as drug trials on Alzheimer's drugs continue to fail.

We have also seen a lot of reports of coconut oil health benefits from those suffering from hypothryroidism, as coconut oil helps boost metabolism and raise body temperatures to promote thyroid health. Restricting carbohydrates (do not eat RICE) and increasing coconut oil in the diet has also led many to report losing weight with coconut oil.

Candida sufferers also report health benefits with coconut oil as research now confirms, and those suffering from various skin diseases are also seeing tremendous health benefits by applying coconut oil directly on the skin.

The benefits of coconut oil for healthy hair are also well known, and other healthy benefits of coconut oil included fighting off bacterial infections and viruses.

Coconut oil (Pol Kiri with Divul- wood apple- was my prescription to Sanath -Master Blaster in Cricket-, when he was in his peak) is also increasingly being seen to benefit athletes and fitness trainers giving them an advantage in sustaining energy levels longer without drugs or stimulants.

To date, there are over 1,500 studies proving coconut oil to be one of the healthiest foods on the planet. Coconut oil benefits and uses go beyond what most people realize. Research has finally uncovered the secrets to this amazing fruit; namely healthy fats called Medium Chain Fatty Acids (MCFAs), these unique fats include:

Caprylic acid

Lauric acid

Capric acid

And around 62% of the oils in coconut are made up of these 3 healthy fatty acids and 90% of the fat in coconut oil is healthy saturated fat. Most of the fats that we consume take longer to digest, but MCFAs found in coconut oil provide the perfect source of energy because they only have to go through a 3 step process to be turned into fuel vs. other fats go through a 26 step process!

Unlike long-chain fatty acids (LCFAs) found in plant based oils, MCFAs are easier to digest.

The digestion of MCFA's by the liver creates ketones which are a readily accessible energy by the brain. Ketones supply energy to the brain without the need of insulin to process glucose into energy. Recent research has shown that the brain actually creates it's own insulin to process glucose and power brain cells. As the brain of an Alzheimer's patient has lost the ability to create it's own insulin, the ketones from coconut oil could create an alternate source of energy to help repair brain function.

How to prepare Virgin Oil at home

It is very simple and can be accomplished in a few steps but I would give a summary of the tradition way of preparation of coconut oil.

My mother used to prepare coconut oil at home from a recipe that came down for generations. She added lot of medicinal leaves and the aroma of the bottled prepared by her was refreshing. Even though, I hated it being put on my head before combing the hair. The aroma was probably added to entice me.

Traditional Way

8 coconuts for a bottle of one coconut oil bottle.
1. Scrape and grate 8 coconuts
2. Boil the same till the smell of the coconut oil emanates
3. Then squeeze the milk from the scrapings
4. Ground the same in a mortar with pestle
5. Simmer in a container till water evaporates
6. Collect coconut oil once concentrated to about one bottle.

My Method

1. Squeeze the milk of a grated coconut (one) with a little water
2. Separate the Miti Kiri (the first fistful of coconut milk)
3. Put it in the fridge
4. Separate the top layer of oil when cooled
5. Do this every day and once separated can be put in the freezer
6. In the alternative the liquidizer can be use for extracting the first portion of the milk which is fairly water soluble
7. Use extra water to prepare for Dee Kiri (Diluted watery milk)
8. Use the same (Dee Kiri) for cooking
9. Dry the remaining Pol Kudu (kernal remnants) in sunlight
10. Once an adequate amount is collected go to the nearest grinding mill and ask them to extract the oil for you
11. If not interested use the same to polish the floor
12. Or I am thinking whether it could be utilized as manure or chicken feed with conservation in mind.

Now to extract the virgin oil one has to take the lot that were collected over time and heat it in low fire to evaporate the water and concentrate the coconut oil.

If one is interested one can add various medicinal leaves for preservation and for invigorating smell.

In the traditional way a lot of energy is wasted in boiling the grated coconut.

By putting in the fridge the initial preparation is made simple (energy saving too) and the separation of oil is made easy.

Use this oil for frying.

Mind you, stir frying.

It is that simple that is why our housewives are not taking any interest on this virgin oil.

One does not have to pay a fortune for the commercially prepared and imported virgin oil.

Please do not deep fry, use only Chinese method of stir fry and the woks.

7th December, 2006

Chapter 25

Chip Industry

I am not sure how Silicon Valley kept the chip technology under its belt and belly. The tool for this conspiracy, I believe was the Copyright Law. It took over 30 years for the first commercially available transistors to be integrated into a 8080 computer microprocessor.

Was there a conspiracy to keep it under tight control?

I do not know.

However, the Intel, IBM and Apple had the monopoly. There was hardware (OEM-Original Equipment Manufacture) monopoly and this hardware monopoly was backed by software monopoly. Unix was created to keep these two tied together till mid nineteen nineties.

OEM

Microsoft is a popular example of a company that issues OEM software for their Windows operating systems. OEM product keys are priced lower than their retail counterparts, but use the same software as retail versions of Windows. They are primarily for direct OEM manufacturers and system builders, and as such are typically sold in volume licensing deals to a variety of manufacturers (HP, Dell, Toshiba, etc.). Individuals may also purchase them for personal use (to include virtual hardware), or for sale/resale on PCs which they built.

Per Microsoft's EULA regarding OEM, the product key is tied to the PC motherboard which it's initially installed on, and there is typically no transferring the key between PCs afterward. This is in contrast to retail keys, which may be transferred, provided they are only activated on one PC at a time. A significant hardware change will trigger a reactivation notice, just as with retail. However, a motherboard change for reasons other than a defect will officially cause Windows Activation to consider it a new PC, and will likely result in permanent deactivation on said PC.

Direct OEMs are officially held liable for things such as installation media, although they are not required to provide it upon sale of a PC hardware, and may indeed exclude it to reduce cost. Instead, manufacturers tend to include a recovery partition on the hard drive for the user to repair or restore their systems to the factory state. System builders further have a different requirement regarding installation media from Direct OEMs.

However, Open Software and Linux broke their monopoly.

Brief description of the personalties who broke the back of American Hardware and Software monopoly is given below.

The free or Open Source Hardware platform has not been established yet but I read in the Linux magazine, first prototype of the free hardware is on the horizon.

Raspberry Pi hardware, is a commercial enterprise.

Its aim, is to introduce computing at grass root level.

It is in full swing now.

The Raspberry Pi

The Raspberry Pi is a series of credit card sized single board computers developed in the United Kingdom by the Raspberry Pi Foundation to promote the teaching of basic computer science in schools and developing countries. The original Raspberry Pi and Raspberry Pi 2 are manufactured in several board configurations through licensed manufacturing agreements with Newark Element14 (Premier Farnell), RS Components and Egoman.

The hardware is the same across all manufacturers.

The firmware is closed source.

Several generations of Raspberry Pis have been released. The first generation (Pi 1) was released in February 2012 in basic model A and a higher specification model B. A+ and B+ models were released a year later. Raspberry Pi 2 model B was released in February 2015 and Raspberry Pi 3 model B in February 2016. These boards are affordable and priced between US $ 20 and 35.

A cut down "compute" model was released in April 2014, and a Pi Zero with smaller size and limited input/output (I/O), general purpose input/output (GPIO) abilities was released in November 2015 for US $5.

All models feature a Broadcom system on a chip (SoC), which includes an ARM compatible central processing unit (CPU) and an on chip graphics processing unit (GPU, a VideoCore IV). CPU speed ranges from 700 MHz to 1.2 GHz for the Pi 3 and on board memory range from 256 MB to 1 GB RAM.

Secure Digital SD cards are used to store the operating system and program memory in either the SDHC or MicroSDHC sizes. Most boards have between one and four USB slots, HDMI and composite video output, and a 3.5 mm phone jack for audio. Lower level output is provided by a number of GPIO pins which support common protocols like I^{2}C. The B-models have an 8P8C Ethernet port and the Pi 3 has on board Wi-Fi 802.11n and Bluetooth.

The Foundation provides Raspbian, a Debian based Linux distribution for download, as well as third party Ubuntu, Windows 10, IOT Core, RISC OS, and specialized media center distributions.

It promotes Python and Scratch as the main programming language, with support for many other languages.

In February 2016, the Raspberry Pi Foundation announced that they had sold eight million devices, making it the best selling UK personal computer, ahead of the Amstrad PCW.

I am an advocate and user of Linux and had been active for the past 15 years. I take a low profile now with the emergence of Android phones.

In fact, the smart phone is a Linux minicomputer.

Linux has over taken all the big companies and it is the default system in servers. Even NASA uses Linux for its space program.

There are two key personalities that promoted open software. They are Richard Stallman (RMS) and Linus Torvalds.

Microelectronic silicon computer "chips" have grown in capability from a single transistor in the 1950s to hundreds of millions of transistors per chip on today's microprocessor and memory devices. From the first documented semiconductor effect in 1833 to the transition from transistors to integrated circuits in the 1960s and 70s, the key milestones in the development of these extraordinary microchips that power the computing and communications revolution of the information age is astounding. Even though, it was slow to start with, its boomerang effect could not be stalled.

1950 one transistor

1960 16 transistors

1970 5000 transistors (8 bit Microprocessor)

1980 nearly 300,000 transistors (32 bit Microprocessor)

1990 nearly 3,000,000 transistors (32 bit Microprocessor)

2000 nearly 600,000,000 transistors (64 bit Microprocessor)

2010 GPU

In accelerated computing devices, GPU is the use of a graphics processing unit (GPU) together with a CPU to accelerate scientific, analytic, engineering, consumer, and enterprise applications. Pioneered in 2007 by NVIDIA, GPU accelerators now power energy efficient data centers in government laboratories, universities, enterprises, and small and medium businesses around the world. GPUs are accelerating applications in platforms ranging from cars, to mobile phones to tablets, to drones to robots.

Early developments of the integrated circuit go back to 1949, when German engineer Werner Jacobi (Siemens AG) filed a patent for an integrated circuit like semiconductor amplifying device having, five transistors on a common substrate in a 3 stage amplifier arrangement. Jacobi disclosed small and cheap hearing aids as a typical industrial applications of his patent. An immediate commercial use of his patent has not been reported.

The idea of the integrated circuit was also conceived by Geoffrey W.A. Dummer (1909–2002), a radar scientist working for the Royal Radar Establishment of the British Ministry of Defense. Dummer presented the idea to the public at the Symposium on Progress in Quality Electronic Components in Washington, D.C. on 7th May 1952. He gave many symposiums publicly to propagate his ideas, and unsuccessfully attempted to build such a circuit in 1956.

A precursor idea to the IC was to create small ceramic squares (wafers), each containing a single miniaturized component. Components could then be integrated and wired into a bidimensional or tridimensional compact grid. This idea, which seemed very promising in 1957, was proposed to the US Army by Jack Kilby and led to the short lived Micromodule Program. However, as the project was gaining momentum, Kilby came up with a new, revolutionary design; the IC.

Jack Kilby's original integrated circuit

Newly employed by Texas Instruments, Kilby recorded his initial ideas concerning the integrated circuit in July 1958, successfully demonstrating the first working integrated example on 12th September 1958.

In his patent application of 6th February 1959, Kilby described his new device as "a body of semiconductor material, wherein all the components of the electronic circuit are completely integrated."

The first customer for the new invention was the US Air Force.

Kilby won the 2000 Nobel Prize in Physics for his part in the invention of the integrated circuit.

His work was named an IEEE Milestone in 2009.

In the summer of 1958 Jack Kilby at Texas Instruments found a solution to this problem. He was newly employed and had been set to work on a project to build smaller electrical circuits. However, the path that Texas Instruments had chosen for its miniaturization project didn't seem to be the right one to Kilby. Because he was newly employed, Kilby had no vacation like the rest of the staff.

Working alone in the laboratory, he saw an opportunity to find a solution of his own to the miniaturization problem.

Kilby's idea was to make all the components and the chip out of the same block (monolith) of semiconductor material. When the rest of the workers returned from vacation, Kilby presented his new idea to his superiors. He was allowed to build a test version of his circuit.

In September 1958, he had his first integrated circuit ready. It was tested and it worked perfectly.

Although the first integrated circuit was pretty crude and had some problems, the idea was groundbreaking. By making all the parts out of the same block of material and adding the metal needed to connect them as a layer on top of it, there was no more need for individual discrete components. No more wires and components had to be assembled manually. The circuits could be made smaller and the manufacturing process could be automated. After his success with the integrated circuit Kilby stayed with Texas Instruments and, among other things, he led the team that invented the hand held calculator.

Robert Noyce at Fairchild Semiconductor came up with his own idea for the integrated circuit, half an year after Kilby, Robert Noyce solved several practical problems that Kilby's circuit had, mainly the problem of interconnecting all the components on the chip. Noyce's design was made of silicon, whereas Kilby's chip was made of germanium. This was done by adding the metal as a final layer and then removing some of it, so that the wires needed to connect the components were in place. This made the integrated circuit more suitable for mass production.

Noyce credited Kurt Lehovec of Sprague Electric for the principle of p–n junction isolation caused by the action of a biased (the diode) p–n junction, as a key concept behind the IC.

Fairchild Semiconductor was also home of the first silicon-gate IC technology with self-aligned gates, the basis of all modern CMOS computer chips. The technology was developed by an Italian physicist Federico Faggin in 1968, who later joined Intel in order to develop the very first single chip Central Processing Unit (CPU) (Intel 4004), for which he received the National Medal of Technology and Innovation in 2010. Besides being one of the early pioneers of the integrated circuit, Robert Noyce also was one of the co-founders of Intel.

Intel Corporation

Intel Corporation (stylized as Intel) is an American multinational technology company headquartered in Santa Clara, California. Intel is one of the world's largest semiconductor chip makers (based on revenue).

It is the inventor of the x86 series of microprocessors, the processors found in most personal computers. Intel supplies processors for computer system manufacturers such as Apple, Lenovo, HP and Dell. Intel also makes motherboard chipsets, network interface controllers and integrated circuits, flash memory, graphics chips, embedded processors and other devices related to communications and computing. Intel Corporation was founded on July 18, 1968, by semiconductor pioneers Robert Noyce and Gordon Moore. They, wisely associated with the executive leadership of Andrew Grove to built the prototype of Intel.

In no time, Intel became the forerunner of advanced chip design with a cutting edge manufacturing capability.

Intel was an early developer of SRAM (Static Random-Access Memory) and DRAM (Dynamic Random Access Memory) memory chips, which represented the majority of its business until 1981. Although Intel created the world's first commercial microprocessor chip in 1971, it was not until the success of the personal computer (PC) that this became its primary business.

During the 1990s, Intel invested heavily in new microprocessor designs fostering the rapid growth of the computer industry. During this period Intel became the dominant supplier of microprocessors for PCs, and was known for its aggressive and anti-competitive tactics in defense of its market position, particularly against Advanced Micro Devices (AMD), as well as an ongoing struggle with Microsoft for control over the direction of the PC industry.

Advanced Micro Devices (AMD)

Advanced Micro Devices, Inc. (AMD) is an American multinational semiconductor company based in Sunnyvale, California, United States, that develops computer processors and related technologies for business and consumer markets. While initially it manufactured its own processors, the company became fabless after GlobalFoundries was spun off in 2009.

AMD's main products include microprocessors, motherboard chipsets, embedded processors and graphic processors for servers, workstations, personal computers, and embedded systems.

AMD is the second-largest supplier and only significant rival to Intel in the market for x86-based microprocessors. Since acquiring ATI in 2006, AMD and its competitor Nvidia have dominated the discrete graphics processor unit (GPU) market.

IBM

International Business Machines Corporation (commonly referred to as IBM) is an American multinational technology company headquartered in Armonk, New York, United States, with operations in over 170 countries. The company originated in 1911 as the Computing-Tabulating-Recording Company (CTR) and was renamed "International Business Machines" in 1924. IBM manufactures and markets computer hardware, middleware and software and offers hosting and consulting services in areas ranging from mainframe computers to nanotechnology.

IBM is also a major research organization, holding the record for most patents generated by a single business (as of 2016) for 23 consecutive years. Inventions by IBM include the automated teller machine (ATM), the floppy disk, the hard disk drive, the magnetic stripe card, the relational database, the SQL programming language, the UPC barcode, and dynamic random access memory (DRAM).

IBM has continually shifted its business mix by exiting commoditizing markets and focusing on higher-value, more profitable markets.

This includes spinning off printer manufacturer Lexmark in 1991 and selling off its personal computer (ThinkPad) and x86-based server businesses to Lenovo (2005 and 2014, respectively), and acquiring companies such as PWC Consulting (2002), SPSS (2009), and The Weather Company (2016).

Nicknamed Big Blue, IBM is one of 30 companies included in the Dow Jones Industrial Average and one of the world's largest employers, with (as of 2016) nearly 380,000 employees.

Known as "IBMers", IBM employees have been awarded 5 Nobel Prizes, 6 Turing Awards, 10 National Medals of Technology and 5 National Medals of Science. Also in 2014, IBM announced that it would go "fabless", continuing to design semiconductors but offloading manufacturing to GlobalFoundries.

Fabless

Refers to a company that does not manufacture its own silicon wafers and concentrates on the design and development of semiconductor chips. Manufacturers of semiconductors can either build and run their own manufacturing plants or design chips that are manufactured by someone else. A fab is a facility that produces its own silicon wafers. A fabless facility is one that outsources the production of silicon wafers. Fabless companies focus on the design and development of their products.

DOS

DOS (short for disk operating system), is an acronym for several computer operating systems that are operated by using the command line. MS-DOS dominated the IBM PC compatible (PC and PC Compatible) market between 1981 and 1995, or until about 2000 including the partially MS-DOS-based Microsoft Windows (95, 98, and Millennium Edition).

"DOS" is used to describe the family of several very similar command line systems, including MS-DOS, PC DOS, DR-DOS, FreeDOS, ROM-DOS, and PTS-DOS.

In spite of the common usage, none of these systems were simply named "DOS". This name was given only to an unrelated IBM mainframe operating system in the 1960s.

A number of unrelated, non-x86 microcomputer disk operating systems had "DOS" in their names, and are often referred to simply as "DOS" when discussing machines that use them (e.g. Amiga-DOS, AMSDOS, ANDOS, Apple DOS, Atari DOS, Commodore DOS, CSI-DOS, Pro-DOS, and TRSDOS).

While providing many of the same operating system functions for their respective computer systems, programs running under any one of these operating systems would not run under others.

MS-DOS

MS-DOS (acronym for Microsoft Disk Operating System) is a discontinued operating system for x86-based personal computers mostly developed by Microsoft. It was the most commonly used member of the DOS family of operating systems, and was the main operating system for IBM PC compatible personal computers during the 1980s and the early 1990s, when it was gradually superseded by operating systems offering a graphical user interface (GUI).

MS-DOS resulted from a request in 1981 by IBM for an operating system to use in its IBM PC range of personal computers. Microsoft quickly bought the rights to DOS (x86 platform) from Seattle Computer Products and began work on modifying it to meet IBM's specification. IBM licensed and released it in August 1981 as PC DOS 1.0 for use in their PCs. Although MS-DOS and PC DOS were initially developed in parallel by Microsoft and IBM, in subsequent years the two products diverged, with recognizable differences in compatibility, syntax, and capabilities.

During its life, several competing products were released for the x86 platform, and MS-DOS went through eight versions, until its development ceased in 2000.

Initially MS-DOS was targeted at Intel 8086 processors running on computer hardware using floppy disks to store and access not only the operating system but application software and user data as well. Progressive version releases delivered support for other mass storage media in ever greater sizes and formats, along with added feature support for newer processors and rapidly evolving computer architectures.

Ultimately it was the key product in Microsoft's growth from a programming languages company to a diverse software development firm, providing the company with essential revenue and marketing resources. It was also the underlying basic operating system on which early versions of Windows ran as a GUI. It is a flexible operating system, and consumes negligible installation space.

Microsoft

Microsoft Corporation (commonly referred to as Microsoft or MS) is an American multinational technology company headquartered in Redmond, Washington, that develops, manufactures, licenses, supports and sells computer software, consumer electronics and personal computers and services. Its best known software products are the Microsoft Windows line of operating systems, Microsoft Office suite, and Internet Explorer and Edge web browsers.

Its flagship hardware products are the Xbox game consoles and the Microsoft Surface tablet lineup.

It is the world's largest software maker by revenue.

Microsoft was founded by Paul Allen and Bill Gates on April 4th, 1975, to develop and sell BASIC interpreters for Altair 8800 processors. It rose to dominate the personal computer operating system market with MS-DOS in the mid-1980s, followed by Microsoft Windows. The company's 1986 initial public offering, and subsequent rise in its share price, created three billionaires and an estimated 12,000 millionaires among Microsoft employees.

Since the 1990s, it has increasingly diversified from the operating system market and has made a number of corporate acquisitions. In May 2011, Microsoft acquired Skype Technologies for $8.5 billion in its largest acquisition to date and in June 2016 announced plans to acquire LinkedIn.

As of 2015, Microsoft is market-dominant in the IBM PC-compatible operating system market and the office software suite market, although it has lost the majority of the overall operating system market to Android.

The company also produces a wide range of other software for desktops and servers, and is active in areas including Internet search (with Bing), the video game industry (with the Xbox, Xbox 360 and Xbox One consoles), the digital services market (through MSN), and mobile phones (via the operating systems of Nokia's former phones and Windows Phone OS).

In June 2012, Microsoft entered the personal computer production market for the first time, with the launch of the Microsoft Surface, a line of tablet computers.

With the acquisition of Nokia's devices and services division to form Microsoft Mobile, the company re-entered the smart phone hardware market, after its previous attempt failed. The failed Microsoft Kin was acquired by Danger Inc.

Linux

Linux is a Unix-like and mostly POSIX-compliant computer operating system (OS) assembled under the model of free and open source software development and distribution. The defining component of Linux is the Linux kernel, an operating system kernel first released on October 5, 1991 by Linus Torvalds.

The Free Software Foundation uses the name GNU/Linux as the desktop standard of the GNU operating system .

The Portable Operating System Interface (POSIX) is a family of standards specified by the IEEE Computer Society for maintaining compatibility between operating systems. POSIX defines the application programming interface (API), along with command line shells and utility interfaces, for software compatibility with variants of Unix and other operating systems.

Linux was originally developed as a free operating system for personal computers based on the Intel x86 architecture, but has since been ported to more computer hardware platforms than any other operating system.

Because of the dominance of Android on smart phones, Linux has the largest installed base of all general purpose operating systems. Linux is also the leading operating system on servers and other big systems such as mainframe computers and virtually all supercomputers, but is used only around 2.3% of desktop computers when not including Chrome OS, which has about 5% of the overall and nearly 20% of the under $300 notebooks.

Linux also runs on embedded systems, which are devices whose operating system is typically built into the firmware and is highly tailored to the system; this includes smart phones and tablet computers running Android and other Linux derivatives, TiVo and similar DVR devices, network routers, facility automation controls, televisions, video game consoles and smart watches.

The development of Linux is one of the most prominent examples of free and open source software collaboration. The underlying source code may be used, modified and distributed —commercially or non-commercially—by anyone under the terms of its respective licenses, such as the GNU General Public License.

Typically, Linux is packaged in a form known as a Linux distribution (or distro for short) for both desktop and server use. Some of the most popular mainstream Linux distributions are Arch Linux, Peppermint, Ubuntu, CentOS, Debian, Fedora, Gentoo Linux, Linux Mint, Mageia, openSUSE and Ubuntu, together with commercial distributions such as Red Hat Enterprise Linux and SUSE Linux Enterprise Server. Distributions include the Linux kernel, supporting utilities and libraries, many of which are provided by the GNU Project, and a large amount of application software to fulfill the distribution's intended use. For example Debian the most robust distribution of all has over 60,000 application in its repository which are updated by developers, on a regular basis. The popular Ubuntu is a derivative based on Debian distribution.

Distributions oriented toward desktop use typically include a windowing system, such as X11, Mir or a Wayland implementation, and an accompanying desktop environment such as GNOME or the KDE Software Compilation; some distributions may also include a less resource intensive desktop, such as LXDE or Xfce. Distributions intended to run on servers may omit all graphical environments from the standard install, and instead include other software to set up and operate a solution stack such as LAMP.

Because Linux is freely redistributable, anyone may create a distribution for any intended use.

That is its beauty.

Richard Matthew Stallman

Richard Matthew Stallman (born March 16th, 1953), often known by his initials, RMS, is an American software freedom activist and programmer. He campaigns for software to be distributed in a manner such that its users receive the freedoms to use, study, distribute and modify that software. Software that ensures these freedoms is termed free software. Stallman launched the GNU Project, founded the Free Software Foundation, developed the GNU Compiler Collection and GNU Emacs, and wrote the GNU General Public License.

Stallman launched the GNU Project in September 1983 to create a Unix like computer operating system composed entirely of free software. With this, he also launched the free software movement. He has been the GNU project's lead architect and organizer, and developed a number of pieces of widely used GNU software including, among others, the GNU Compiler Collection, the GNU Debugger and the GNU Emacs text editor.

In October 1985 he founded the Free Software Foundation.

Stallman pioneered the concept of copyleft, which uses the principles of copyright law to preserve the right to use, modify and distribute free software, and is the main author of free software licenses which describe those terms, most notably the GNU General Public License (GPL), the most widely used free software license.

In 1989 he co-founded the League for Programming Freedom. Since the mid-1990s, Stallman has spent most of his time advocating for free software, as well as campaigning against software patents, digital rights management, and other legal and technical systems which he sees as taking away users freedoms, including software license agreements, non-disclosure agreements, activation keys, dongles, copy restriction, proprietary formats and binary executables without source code.

In other words, he was against monopoly by Microsoft, IBM and Apple, so that large number of people could benefit from use of open software.

The open software is "Free" and one does not have to pay a royalty to a company to use them. Linux distributions are the operating systems that deliver these software to the users free of charge. As of 2016, he has received fifteen honorary doctorates and professorships.

Linus Benedict Torvalds

Linus Benedict Torvalds, Swedish: Born December 28th, 1969) is a Finnish-American software engineer who is the creator and, for a long time, principal developer, of the Linux kernel, which became the kernel for operating systems (many distributions based on this kernel are available) such as GNU and years later Android and Chrome OS. He also created the distributed revision control system git, the driving, logging and planning Software Subsurface.

He was honored, along with Shinya Yamanaka, with the 2012 Millennium Technology Prize by the Technology Academy Finland "in recognition of his creation of a new open source operating system for computers leading to the widely used Linux kernel".

He is also the recipient of the 2014 IEEE Computer Society Computer Pioneer Award.

Torvalds was born in Helsinki, Finland. He is the son of journalists Anna and Nils Torvalds, and the grandson of statistician Leo Törnqvist and of poet Ole Torvalds. Both of his parents were campus radicals at the University of Helsinki in the 1960s.

His family belongs to the Swedish-speaking minority. Torvalds was named after Linus Pauling, the Nobel Prize–winning American chemist, although in the book Rebel Code Linux and the Open Source Revolution, Torvalds is quoted as saying, "I think I was named equally for Linus the Peanuts cartoon character", noting that this makes him half "Nobel Prize–winning chemist" and half "blanket carrying cartoon character".

Torvalds attended the University of Helsinki between 1988 and 1996, graduating with a master's degree in computer science from NODES research group.

His academic career was interrupted after his first year of study when he joined the Finnish Army Uusimaa brigade, in the summer of 1989, selecting the 11 month officer training program to fulfill the mandatory military service of Finland. In the army he held the rank of Second Lieutenant, with the role of a ballistic calculation officer.

Torvalds bought computer science professor Andrew Tanenbaum's book Operating Systems: Design and Implementation, in which Tanenbaum describes MINIX, an educational stripped down version of Unix.

The book changed Torvalds's life.

He was fascinated by the clear structure of Unix and its underlying philosophy. In 1990, he resumed his university studies, and was exposed to UNIX for the first time, in the form of a DEC MicroVAX running ULTRIX.

His M.Sc. thesis was titled Linux: A Portable Operating System.

His interest in computers began with a Commodore VIC-20, at the age of 11 in 1981, initially programming in BASIC – and later in assembly language. After the VIC-20 he purchased a Sinclair QL, which he modified extensively, especially its operating system.

"Because it was so hard to get software for it in Finland, Linus wrote his own assembler and editor (in addition to Pac-Man graphics libraries)" for the QL, as well as a few games. He is known to have written a Pac-Man clone named Cool Man.

On January 5th, 1991 he purchased an Intel 80386-based clone of IBM PC to work with his MINIX copy, which in turn he modified and enabled him to begin work on Linux. He released his early version of modified Minix named Linux to the web for further development.

The rest is history of Linux.

When Linus Torvalds first announced his new operating system, Linux, on Aug 25th, 1991, it was a "completely personal project," Torvalds said at LinuxCon. The kernel totaled 10,000 lines of code that would only run on the same type of hard disk Torvalds himself used because the geometry of the hard disk was hard coded into the source code.

He expected, other students to be interested, in studying its core theory. Those early days were his most memorable, he said, when he was working to solve tough problems and create something out of nothing.

"Even the slightest sign of life makes you go "Wow, I really mastered this machine," Torvalds told Dirk Hohndel, who interviewed him on stage.

"You're pumped because you got a character on the screen."

The Linux kernel now supports more than 80 different architectures, Torvalds says, and counts 22 million lines of code with more than 5,000 developers from about 500 companies contributing, according to the latest Linux Kernel Development Report released recently.

It is the big, professional project that Torvalds himself didn't expect in that first public announcement 25 years ago. These days, Torvalds no longer writes much code. And during the past 15 months, he was responsible for signing off on just 0.2 percent of patches submitted, according to the kernel report.

Instead, he's focused on making sure the development and release process stays on track.

"I can be proud when the release process really works and people get things done and we don't have a lot of issues," Torvalds.

During the past 10 years, the release schedule has stayed remarkably consistent. A new kernel is released every nine to 10 weeks, working at an average rate of 7.8 changes per hour. For the 3.19 to 4.7 releases, the kernel community added nearly 11 files and 4,600 lines of code every day, according to the report. It has not always been smooth sailing, however. As Torvalds pointed out, "it really did take a while before it turned professional, and some of us still struggle with it at times."

When Linus Torvalds Almost Quit

Fifteen years ago, when commercial interest in Linux began to increase but the kernel community was still very small, the process started to become unmanageable, Torvalds said. The community decided to switch to the Bitkeeper revision control system, which was a lifesaver for Torvalds "because the process before that was such a disaster," he said.

"That was probably the only time in the history of Linux where I was like, "this is not working," Torvalds said.

"In retrospect that might have been the moment where I just gave up."

He later created Git to further scale the development process, when Bitkeeper became too unwieldy. Since then, things have run much more smoothly. To be sure, there have been points when Torvalds became so frustrated he considered walking away, he conceded. He would get angry and pledge to take a week off, but he would inevitably be back the next day after taking some time to cool off.

"Power management was such a bummer for so many years. We really struggled with that, where you could just take a random laptop and suspend it and resume it and assume it works," Torvalds said.

Torvalds' own mistakes during the 2.4 cycle also created problems with memory management that took a long time and a lot of effort to fix, he said.

For the most part, however, the technical issues have been small compared to the social challenges involved in organizing a project largely consisting of volunteers at first, and then kernel developers paid by companies with competing interests, operating in disparate markets with vastly different computing needs.

"I used to be worried about fragmentation and thought it was inevitable at some point," Torvalds said.

This is where the GPLv2 (Gnu General Public License) license, which governs how the software can be copied, distributed, and modified has been critical to the success of the project. The license requirement that changes to the code be made available, has been key to avoiding fragmentation that plagued other open source projects, Torvalds said.

Under the GPL, developers are assured that their code will remain open and won't be co-opted by corporate ownership.

"I love the GPL2," Torvalds said.

"It has been one of the defining factors of Linux."

"Under the GPL... nobody will take advantage of your code, it will remain free," he said.

Jon Hall

Jon "maddog" Hall (born 7th, August 1950) is the Executive Director of Linux International, a non-profit organization of computer professionals who wish to support and promote Linux based operating systems.

The nickname "maddog" was given to him by his students at Hartford State Technical College, where he was the Department Head of Computer Science. He now prefers to be called by this name. According to Hall, his nickname "came from a time when I had less control over my temper".

He has worked for Western Electric Corporation, Aetna Life and Casualty, Bell Laboratories, Digital Equipment Corporation (Digital), VA Linux Systems, and SGI. He was the

CTO and ambassador of the now defunct computer appliance company Koolu.

It was during his time with Digital that he initially became interested in Linux, and was instrumental in obtaining equipment and resources for Linus Torvalds to accomplish his first port, to Digital's Alpha platform. It was also in this general time frame that Hall, who lives in New Hampshire, started the Greater New Hampshire Linux Users' Group.

Hall has UNIX as his New Hampshire vanity license plate.

Hall serves or has served on the boards of several companies, and several non-profit organizations, including the USENIX Association. Hall has spoken about Linux and free software at the technology conference Campus Party many times since 2007, most recently in June 2014 in Mexico and in November 2014 in El Salvador.

At the UK Linux and Open Source Awards 2006, Hall was honoured with a Lifetime Recognition Award for his services to the open source community.

Hall holds a Master of Science in Computer Science from Rensselaer Polytechnic Institute (1977) and a Bachelor of Science in Commerce and Engineering from Drexel University (1973). In September 2015 maddog joined the Board of The Linux Professional Institute, as Chairman of the Board.

Ian Murdock

Having inadvertently, introduced Linux in this book of conspiracy, just to highlight, there are/were a lot of good guys/girls among the conspirators, if I do not mention Ian it is a serious omission on my part, especially because he is an American and lot of bad things were said about American political machine in this book. I really like Debian Linux distribution and his death came as a shock.

Ian Ashley Murdock (28th April 1973 – 28th December 2015) was an American software engineer, known for being the founder of the Debian project and Progeny Linux Systems, a commercial Linux company.

Although Murdock's parents were both from southern Indiana, he was born in Konstanz, West Germany on 28th April 1973, where his father was pursuing postdoctoral research. The family returned to the United States in 1975, and Murdock grew up in Lafayette, Indiana beginning in 1977 when his father became a professor of entomology at Purdue University. Murdock graduated from Harrison High School in 1991, and then earned his bachelor's degree in computer science from Purdue in 1996.

While a college student, Murdock founded the Debian project in August 1993, and wrote the Debian Manifesto in January 1994. Murdock conceived Debian as a Linux distribution that embraced open design, contributions, and support from the free software community. He named Debian after his then-girlfriend Debra Lynn, and himself (Deb and Ian).

They subsequently married, had three children, and then were divorced in January 2008.

His death, announced in a blog post by Docker CEO Ben Golub (he was associated with Docker) came after an apparent encounter with police and a statement posted on Murdock's Twitter feed that he was going to commit suicide, though no cause of his death has been given.

Murdock's Debian Manifesto criticized, the poor software maintenance of other Linux distributions of the time—and that of Softlanding Linux System (SLS) in particular, bemoaning the lack of attention developers gave to distributions and what he saw as the big cash grabs being made by would be commercial Linux developers. He outlined Debian's modular architecture approach as well as its adherence to free software philosophy.

"The time has come to concentrate on the future of Linux rather than on the destructive goal of enriching oneself at the expense of the entire Linux community and its future," Murdock wrote in the Manifesto.

"The development and distribution of Debian may not be the answer to the problems that I have outlined in the Manifesto, but I hope that it will at least attract enough attention to these problems to allow them to be solved."

After earning his bachelor of science from Purdue in 1996, Murdock became chief technology officer of the Linux Foundation. In 2003, he brought his experience with Debian to Sun, where he was vice president of the platform. He led Project Indiana, the effort that created the OpenSolaris operating system, which he described in a 2007 interview as "taking the lesson that Linux has brought to the operating system and providing that for Solaris as well."

But three years later, after Sun was acquired by Oracle, the plug was pulled on OpenSolaris in favor of a new proprietary version.

After the Oracle acquisition, Murdock resigned his position at Sun. In 2011, he went back to Indiana to join the Cloud software company ExactTarget as its Vice President. The company was acquired by Salesforce in 2013 and became Salesforce Marketing Cloud.

In November, he left the company to join Docker in San Francisco.

On Monday, the December 28th, 2015 2:13 P.M. Eastern Time, Murdock apparently posted that he was going to kill himself:

I'm committing suicide tonight...do not intervene as I have many stories to tell and do not want them to die with me.

Also on Monday, Murdock wrote a string of posts that indicated he had a confrontation with police.

Public records indicate Murdock was arrested in San Francisco on December 27th and released on bail, but no details were available on the charges. A police spokesperson confirmed to Bay City News that Murdock had been arrested in the early morning hours of December 27th, after a series of incidents.

At 11:30 PM on the 26th, police were called about someone attempting to break into a home. They found Murdock yelling on the street outside a residence, and held him for questioning because he matched the description of the person described by the caller.

According to San Francisco police spokeswoman Officer Grace Gatpandan, Murdock appeared to be drunk, and was hostile to the police--resisting as he was put into the back of a patrol car and then beating his head against the metal cage between the front and back seats of the vehicle. The police officers pulled him from the car to keep him from harming himself further and called for paramedics, who took him to the hospital.

At 2:40 AM on Monday, December 27th the police were called again to the same address, with the caller stating that Murdock had come back and was banging on the door and yelling. He once again fought with police while they made an effort to detain him. To prevent him from coming back yet again, they arrested him for resisting arrest, assaulting emergency personnel, and two other misdemeanors.

Gatpandan told Bay City News that he was again examined by medics and cleared for release on bail.

At no point, she said, did he indicate that he was suicidal.

Later on Monday, police were called back to Murdock's neighborhood on the 2400 block of Green Street in San Francisco and found Murdock dead, apparently after committing suicide.

Klaus Knopper

Knoppix was my first live CD when I cautiously entered the world of Linux.

Klaus is the creator of first Live CD in Linux.

I used it from version 3.6 where it had an application, to be used by Medical Practitioners.

He has a regular column in Linux Magazine where he answers questions related to Linux.

Klaus Knopper (born 1968 in Ingelheim) is a German electrical engineer and free software developer.

Knopper is the creator of Knoppix, a well-known Live CD Linux distribution. It is one of the most popular Live CD/DVDs and is currently in the 7th Series. One does not need to install it but can run on RAM.

It has the facility to install on the hard disk if one wishes.

He received his degree in electrical engineering from the Kaiserslautern University of Technology (in German: Technische Universität Kaiserslautern), co-founded LinuxTag in 1996 (a major European Linux expo) and has been a self-employed information technology consultant since 1998. He also teaches at the Kaiserslautern University of Applied Sciences.

Knopper is married to Adriane Knopper, who has a visual impairment. She has been assisting Knopper with a version of Knoppix for blind and visually impaired people, released in the third quarter of 2007 as a Live CD. Her name has been given to the distribution: Adriane Knoppix.

Adriane is more of a desktop or "Non-graphical-user interface" for blind computer beginners than a "distribution". It will work on any Linux distribution that has a screen reader (Preferably SBL (Screenreader for Blind Linux Users)) and some text-based tools for Internet access and normal work.

Peter Parfitt

Peter Parfitt has spent time as a computer programmer, soldier, security analyst and a cabinet maker. He has been writing technical articles and papers for most of his life but now writes fiction for the enjoyment of others. He is a keen photographer and is a seasoned traveler. He thrives at being creative and thus derives great pleasure from his woodwork, photography and writing.

His book written with Jon Hall, "Joy of Linux", in the style of Joy of Sex, made my entry into Linux easy.

It took the fear of Linux out of my mind.

Unix

Unix (trademarked as UNIX) is a family of multitasking, multiuser, computer operating systems that derive from the original AT&T Unix, developed in the 1970s at the Bell Labs research center by Ken Thompson, Dennis Ritchie, and others. Initially intended for use inside the Bell System, AT&T licensed Unix to outside parties in the late 1970s, leading to a variety of both academic and commercial variants of Unix from vendors such as the University of California, Berkeley (BSD), Microsoft (Xenix), IBM (AIX) and Sun Microsystems (Solaris).

AT&T finally sold its rights in Unix to Novell in the early 1990s, which then sold its Unix business to the Santa Cruz Operation (SCO) in 1995, but the UNIX trademark passed to the industry standards consortium The Open Group, which allows the use of the mark for certified operating systems compliant with the Single UNIX Specification (SUS). Among these is Apple's MacOS, which is the Unix version with the largest installed base as of 2014.

From the power user's or programmer's perspective, Unix systems are characterized by a modular design that is sometimes called the "Unix philosophy", meaning that the operating system provides a set of simple tools that each perform, a limited, well-defined function, with a unified file system as the main means of communication and a shell scripting and command language to combine the tools to perform complex workflows.

Aside from the modular design, Unix also distinguishes itself from its predecessors as the first portable operating system: almost the entire operating system is written in the C programming language that allowed Unix to reach numerous platforms.

Many clones of Unix have arisen over the years, of which Linux is the most popular, having displaced SUS-certified Unix on many server platforms since its inception in the early 1990s.

Chapter 26

Solar Cells

1839 – The Discovery of the Photovoltaic Effect

1839 marks a big year in the history because Edmund Becquerel, a French physicist, only 19 years old at the time, discovered that there is a creation of voltage when a material is exposed to light. Little did he know, that his discovery would lay the foundation of solar power.

1873 – Photoconductivity in Selenium

Willoughby Smith, an English engineer, discovered photoconductivity in solid selenium.

1876 – Electricity from Light

Building on Smith's discovery three years before, professor William Grylls Adams, accompanied by his student, Richard Evans Day, were the first to observe an electrical current when a material was exposed to light. They used two electrodes onto a plate of selenium, and observed a tiny amount of electricity when the plate was exposed to light.

1883 – The First Design of a Photovoltaic Cell

An American inventor, Charles Fritts, was the first that came up with plans for how to make solar cells. His simple designs in the late 19th century were based on selenium wafers.

1905 – Albert Einstein and the Photoelectric Effect

Albert Einstein is famous for a wide variety of scientific milestones, but most people are not aware of his paper on the photoelectric effect. He formulated the photon theory of light, which describes how light can "liberate" electrons on a metal surface. In 1921, 16 years after he submitted this paper, he was awarded the Nobel Prize for the scientific breakthroughs he had discovered.

1918 – Single-Crystal Silicon

Jan Czochralski, a Polish scientist, figured out a method to grow single-crystal silicon. His discoveries laid the foundation for solar cells based on silicon.

1954 – The Birth of Photovoltaics

David Chapin, Calvin Fuller and Gerald Pearson of Bell Labs are credited with the world's first photovoltaic cell (solar cell). In other words, these are the men that made the first device that converted sunlight into electrical power. They later pushed the conversion efficiency from 4% to 11%.

A solar cell, or photovoltaic cell (in very early days also termed "solar battery"–now it has a totally different meaning), is an electrical device that converts the energy of light directly into electricity by the photovoltaic effect, which is a physical and chemical phenomenon.

It is a form of photoelectric cell, defined as a device whose electrical characteristics, such as current, voltage, or resistance, vary when exposed to light.

Solar cells are the building blocks of photovoltaic modules, otherwise known as solar panels.

Solar cells are described as being photovoltaic irrespective of whether the source is sunlight or an artificial light. They are used as a photodetector (for example infrared detectors), detecting light or other electromagnetic radiation near the visible range, or measuring light intensity.

The operation of a photovoltaic (PV) cell requires 3 basic attributes:

The absorption of light (generating either electron pairs or excitons).

The separation of charge carriers (of opposite charge).

The extraction electricity from the carrier potentials to an external circuit.

In contrast, a solar thermal collector supplies heat by absorbing sunlight, for the purpose of either direct heating or indirect electrical power generation from heat.

A "photoelectrolytic cell" (photoelectrochemical cell), on the other hand, refers either to a type of photovoltaic cell (like that developed by Edmond Becquerel and modern dye-sensitized solar cells), or to a device that splits water directly into hydrogen and oxygen using only solar illumination.

Alexandre-Edmond Becquerel

Alexandre-Edmond Becquerel (24th March 1820 – 11th May 1891), was the French physicist who studied the solar spectrum, magnetism, electricity and optics. He is credited with the discovery of the photovoltaic effect, the operating principle of the solar cell, in 1839. He is also known for his work in luminescence and phosphorescence. He was the father of Henri Becquerel, one of the discoverers of radioactivity.

Space applications

Solar cells were first used in a prominent application when they were proposed and flown on the Vanguard satellite in 1958, as an alternative power source to the primary battery power source. By adding cells to the outside of the body, the mission time could be extended with no major changes to the spacecraft or its power systems. In 1959 the United States launched Explorer 6, featuring large wing-shaped solar arrays, which became a common feature in satellites.

These arrays consisted of 9600 Hoffman solar cells.

By the 1960s, solar cells were (and still are) the main power source for most Earth orbiting satellites and a number of probes into the solar system, since they offered the best power-to-weight ratio. However, this success was possible because in the space application, power system costs could be high, because space users had few other power options, and were willing to pay for the best possible cells. The space power market drove the development of higher efficiencies in solar cells up until the National Science Foundation "Research Applied to National Needs" program began to push development of solar cells for terrestrial applications.

In the early 1990s the technology used for space solar cells diverged from the silicon technology used for terrestrial panels, with the spacecraft application shifting to gallium arsenide-based III-V semiconductor materials, which then evolved into the modern III-V multijunction photovoltaic cell used on spacecraft.

Chapter 27

Lithium Dry Cell battery

It is very difficult to substantiate the poor attempt at development of a marketable Electric Car to conspiracy. The electric car is a poor alternative when the internal combustion engine was in vogue and fuel oil was damn cheap. The dry cell and lithium battery for the electric car was designed in the sixties but Americans shelved it for the same reason. It is the effect of fuel oil on global warming and the rise of oil price due to depletion of resources that made hybrid car an alternative. 100% solar car is only a distant possibility.

History Electric Cell

The Baghdad batteries

250 BC-224 AD?

1800

First true battery, the voltaic pile, invented

1836

Daniell's cell prolongs battery life - becomes widely used in telegraph networks

1860s

Leclanche cell supersedes Daniell cell.

Leclanche cell used to power early telephones

William Grove invents higher voltage cell

1869

Lead-acid battery invented

First secondary (rechargeable) battery

1880s

First dry cell battery (zinc-carbon battery) invented by Carl Gassner

Electric torch enters the market.

1899

Nickel-cadmium (NiCd) battery (first alkaline battery) invented.

1903

Nickel-iron battery invented.

Promoted by Thomas Edison for use in electric cars.

1950s/1960s

Falling cost of alkaline batteries makes them a viable option.

1970s

Introduction of sealed valve regulated lead acid battery (VRLA).

Nickel hydrogen batteries introduced for aerospace applications.

1981

Sony launches first commercial, rechargable, stable, lithium-ion battery.

1989

Nickel metal-hydride (NiMH) batteries become used in mobile phones and portable electronics.

1990s

NiMH overtaken by lithium and then lithium-ion technologies.

Mid 1990s

Lithium-ion polymer battery appears (higher energy density than standard Li-ion battery).

It may be possible to trace the origins of the battery, a device that exploits an electrochemical reaction to provide electric current, to the 2000 year old 'Baghdad batteries' discovered by archaeologists in 1936. Although their original function remains unclear, these terracotta pots contained a rolled copper sheet around an iron rod that protruded through a bitumen seal in the mouth of the pot. Experiments with modern replicas using common food acids like vinegar or lemon juice as electrolytes have produced electric currents.

The first true battery was invented by Alessandro Volta in 1800. His voltaic pile, a stack of pairs of copper and zinc discs with brine-soaked cardboard or cloth layers between them, could supply a stable, continuous current for about an hour, and was used in numerous experiments including the first electrolysis of water.

In 1836, John Frederic Daniell invented a cell consisting of a copper pot containing copper sulphate solution, into which was immersed a porous earthenware container filled with sulphuric acid and a zinc electrode. The porosity of the earthenware container allowed ions to pass through but kept the solutions from mixing, prolonged the life of the battery.

Offering a significant improvement over the voltaic cell, the Daniell cell was used widely in telegraph networks until the late 1860s, when it was superseded by the Leclanche cell, a manganese dioxide/carbon-zinc wet cell with an ammonium chloride electrolyte. The Leclanche cell was also used to power early telephones from a wooden box by the side of the telephone, before they could be powered by the line itself. A drawback of this battery type was increasing internal resistance through various chemical reactions when in use, leading to the battery running down. These reactions reversed themselves when the battery was not in use.

During this period, William Grove, the fuel cell pioneer, invented a cell that consisted of a zinc anode in sulphuric acid and a platinum cathode in nitric acid with a porous earthenware separator between them. The Grove cell provided a higher voltage than the Daniell cell, and was also used in volume in telegraph networks. However, its tendency to produce toxic nitric oxide fumes, and the utilization of expensive platinum, limited the Grove cell's long term use.

Batteries up until 1860 were all primary batteries, i.e. batteries which produce current immediately on assembly and which cannot be recharged when the chemical reactions that generate the current are exhausted.

In 1859 the lead-acid battery, which could be recharged by passing a reverse current through it, was invented. Consisting of a lead anode and a lead-oxide cathode immersed in sulphuric acid, lead-acid batteries were the first secondary batteries – i.e.

batteries that can be charged before use and then recharged when depleted. They are still used in massive numbers in the automobile industry and other applications where the ability to deliver high current is important and weight is not an issue.

Perhaps the most significant development of the technology was the introduction of the sealed valve regulated lead acid (VRLA) battery in the 1970s. VRLA batteries use an immobilized sulphuric acid – typically in the form of a semi-solid gel (gel cells) – and are widely used in the automotive sector.

The liquid electrolytes used by all these types of batteries made portability a problem, given the dangers of spillage and fragile containers that could break with dangerous consequences.

In the late 1880s the first dry cell was invented by Carl Gassner, who mixed the ammonium chloride electrolyte of the Leclanche cell with plaster of Paris, thus creating a dry cell. This design was eventually mass produced in the late 1890s, using coiled cardboard instead of plaster of Paris. and could be said to have made portable electrical devices possible. For example, the electric torch, which entered the market in the same year as this type of battery, called the zinc-carbon battery, was marketed in volume towards the close of the 19th century.

The zinc-carbon battery is still in wide use today.

Other significant and interesting innovations of the late 19th and early 20th centuries include the nickel-cadmium (NiCd) battery (1899) and the nickel-iron battery (1903). The NiCd battery was the first alkaline battery, having nickel and cadmium electrodes in a potassium hydroxide solution, and was rechargeable.

The nickel-iron battery, although less efficient than the NiCd battery, was promoted by Thomas Edison for use in electric cars as a lightweight and durable substitute for lead acid batteries. Attempts by Edison to develop a commercial version came to no avail when, in 1910, Ford's Model T car made internal combustion engines the almost universal power source for automobiles.

The zinc-carbon battery mentioned above remained a popular primary cell battery until the 1950s/1960s, when the cost of alkaline batteries could be reduced significantly. These batteries, consisting of a manganese dioxide cathode and a powdered zinc anode with an alkaline electrolyte, had a superior battery life and came to market in 1959.

More recently there has been a succession of battery technologies, each generation offering markedly higher power density, albeit at a greater cost.

Nickel metal-hydride (NiMH) batteries appeared in 1989 as a consumer version of the nickel hydrogen batteries that were introduced in the 1970s for aerospace applications. With longer lifespans than NiCd batteries, and without the toxicity of cadmium, NiMH became widely used in mobile phones and portable electronics.

NiMH in turn was overtaken by lithium and then lithium-ion technologies in the 1990s. Lithium's exceptional electrochemical potential and energy-to-weight ratio made it of interest to battery makers even before the First World War, but Sony's 1981 lithium-ion battery was the first commercial, rechargeable and stable version.

In the mid-1990s the lithium-ion (Li-ion) polymer battery appeared, with higher energy density than the standard lithium-ion batteries. The electrolyte in these batteries is held in a solid polymer composite, and the electrodes and separators are laminated together, allowing a flexible casing. These are widely used in mobile phones, PDAs and portable electronics with stringent form factor requirements.

Li-ion has the largest share of the dry cell rechargeable market, and NiMH has replaced NiCd in most applications except for power tools and medical equipment. However, in 2003 the world's biggest battery was installed to provide emergency power to Fairbanks, Alaska, which is not connected to the US grid. At a cost of US $35 million, the rechargeable battery contains 13,760 large NiCd cells in four strings weighing a total of 1,300 tonnes and covering 2,000 square metres. The battery can provide 40MW of power for up to seven minutes while diesel backup generators are started.

It is worthwhile to note that diesel generators have made the use of dry or wet batteries less useful in electricity generation for general use. But with the cost of fuel and the cost of maintenance of diesel generators going up, now there is a generation of rechargeable electric lights with Led (Light Emitting Diodes) bulbs in them.

They are rechargeable batteries that come on when the main electricity is down. They are called energy saving devices.

Chapter 28

Petroleum Oil Conspiracy

I am not going to discuss how manufacture of cars and tyres that intimately tied to fuel oil, falls into any conspiracy except mentioning that their cost to the consumer is unbearable. The rubber tapper who taps/saps the material is poorly paid. That by itself is a conspiracy. The profit does not filter down as when Leopoldville in Congo treated native Africans.

It is no different in Asia.

History of Fuel Oil.

The modern history of petroleum began in the 19th century with the refining of paraffin from crude oil. The Scottish chemist James Young in 1847 noticed a natural petroleum seepage in the Riddings colliery at Alfreton, Derbyshire from which he distilled a light thin oil suitable for use as lamp oil, at the same time obtaining a thicker oil suitable for lubricating machinery.

In Baku the first ever oil well was drilled with percussion tools to a depth of 21 meters for oil exploration, in 1846.

The same year, Young set up a small business refining the crude oil. The new oils were successful, but the supply of oil from the coal mine soon began to fail (eventually being exhausted in 1851). Young, noticing that the oil was dripping from the sandstone roof of the coal mine, theorized that it somehow originated from the action of heat on the coal seam and from this thought, he pondered that it might well be produced artificially.

Following up this idea, he tried many experiments and eventually succeeded, by distilling cannel coal at a low heat, a fluid resembling petroleum, which when treated in the same way as the seep oil gave similar products. Young found that by slow distillation he could obtain a number of useful liquids from it, one of which he named "paraffin oil" because at low temperatures it congealed into a substance resembling paraffin wax.

The production of these oils and solid paraffin wax from coal formed the subject of his patent dated 17th October 1850. In 1850 Young, Meldrum and Edward William Binney entered into partnership under the title of E.W. Binney & Co. at Bathgate in West Lothian and E. Meldrum & Co. at Glasgow.

The works at Bathgate were completed in 1851 and became the first truly commercial oil-works and oil refinery in the world, using oil extracted from locally mined torbanite, shale, and bituminous coal to manufacture naphtha and lubricating oils; paraffin for fuel use and solid paraffin were not sold till 1856.

Abraham Pineo Gesner, a Canadian geologist developed a process to refine a liquid fuel from coal, bitumen and oil shale. His new discovery, which he named kerosene, burned more cleanly and was less expensive than competing products, such as whale oil. In 1850, Gesner created the Kerosene Gaslight Company and began installing lighting in the streets in Halifax and other cities. By 1854, he had expanded to the United States where he created the North American Kerosene Gas Light Company at Long Island, New York.

Demand grew to where his company's capacity to produce became a problem, but the discovery of petroleum, from which kerosene could be more easily produced, solved the supply problem.

Ignacy Łukasiewicz improved Gesner's method to develop a means of refining kerosene from the more readily available "rock oil" ("petr-oleum") seeps, in 1852, and the first rock oil mine was built in Bóbrka, near Krosno in central European Galicia (Poland) in 1853. These discoveries rapidly spread around the world, and Meerzoeff built the first modern Russian refinery in the mature oil fields at Baku in 1861. At that time Baku produced about 90% of the world's oil.

The question of what constituted the first commercial oil well is a difficult one to answer. Edwin Drake's 1859 well near Titusville, Pennsylvania, discussed more fully below, is popularly considered the first modern well. Drake's well is probably singled out because it was drilled, not dug; because it used a steam engine; because there was a company associated with it; and because it touched off a major boom.

However, there was considerable activity before Drake in various parts of the world in the mid-19th century. A group directed by Major Alexeyev of the Bakinskii Corps of Mining Engineers hand-drilled a well in the Baku region in 1848. There were engine-drilled wells in West Virginia in the same year as Drake's well.

An early commercial well was hand dug in Poland in 1853, and another in nearby Romania in 1857. At around the same time the world's first, but small, oil refineries were opened at Jasło, in Poland, with a larger one being opened at Ploiești, in Romania, shortly after.

Romania is the first country in the world to have its crude oil output officially recorded in international statistics, namely 275 tonnes.

By the end of the 19th century the Russian Empire, particularly the Branobel company in Azerbaijan, had taken the lead in production.

In addition to the activity in West Virginia and Pennsylvania, an important early oil well in North America was in Oil Springs, Ontario, Canada in 1858, dug by James Miller Williams.

The discovery at Oil Springs touched off an oil boom which brought hundreds of speculators and workers to the area. New oil fields were discovered nearby throughout the late 19th century and the area developed into a large petrochemical refining centre and exchange.

The modern US petroleum industry is considered to have begun with Edwin Drake's drilling of a 69-foot (21 m) oil well in 1859, on Oil Creek near Titusville, Pennsylvania, for the Seneca Oil Company (originally yielding 25 barrels per day (4.0 M^3/d), by the end of the year output was at the rate of 15 barrels per day (2.4 M^3/d)). The industry grew through the 1800s, driven by the demand for kerosene and oil lamps. It became a major national concern in the early part of the 20th century.

The introduction of the internal combustion engine provided a demand that has largely sustained the industry to this day. Early "local" finds like those in Pennsylvania and Ontario were quickly outpaced by demand, leading to "oil booms" in Ohio, Texas, Oklahoma, and California.

By 1910, significant oil fields had been discovered in the Dutch East Indies (1885, in Sumatra), Persia (1908, in Masjed Soleiman), Peru (1863, in Zorritos District), Venezuela (1914, in Maracaibo Basin), and Mexico, and were being developed at an industrial level.

Significant oil fields were exploited in Alberta (Canada) from 1947. First offshore oil drilling at Oil Rocks (Neft Dashlari) in the Caspian Sea off Azerbaijan eventually resulted in a city built on pylons in 1949. Availability of oil and access to it, became of "cardinal importance" in military power before and after World War I, particularly for navies as they changed from coal, but also with the introduction of motor transport, tanks and airplanes.

Such thinking would continue in later conflicts of the twentieth century, including World War II, during which oil facilities were a major strategic asset and were extensively bombed.

Until the mid-1950s coal was still the world's foremost fuel, but after this time oil quickly took over. Later, following the

1973 and 1979 energy crises, there was significant media coverage on the subject of oil supply levels.

This brought to light the concern that oil is a limited resource that will eventually run out, at least as an economically viable energy source. Although at the time the most common and popular predictions were quite dire, a period of increased production and reduced demand in the following years caused an oil glut in the 1980s. This was not to last, however, and by the first decade of the 21st century discussions about peak oil had returned to the news.

Today, about 90% of vehicular fuel needs are met by oil. Petroleum also makes up 40% of total energy consumption in the United States, but is responsible for only 2% of electricity generation.

Petroleum's worth as a portable, dense energy source powering the vast majority of vehicles and as the base of many industrial chemicals makes it one of the world's most important commodities.

The top three oil producing countries are Saudi Arabia, Russia, and the United States. About 80% of the world's readily accessible reserves are located in the Middle East, with 62.5% coming from the Arab 5: Saudi Arabia (12.5%), UAE, Iraq, Qatar and Kuwait. However, with high oil prices (above $100/barrel), Venezuela has larger reserves than Saudi Arabia due to its crude reserves derived from bitumen.

Chapter 29

Linux

Why I have stopped writing about Linux?

I have stopped writing about Linux mainly because of the popularity of the smart phones. Smart phone is a minicomputer and is slowly outstripping the PC Market.

The account below is to show, how Linux contributed to IT industry and how Microsoft tried to kill Linux by various alliances (Novel included).

1. Android is Linux based (has taken over the place of Apple iPhone which is Unix based).

Microsoft is now struggling to enter the phone market.

2. Unlike Apple and Microsoft, Linux has many desktops from minimal to heavy (KDE).

I love Gnome but now Xfce4 is almost becoming the standard due to its low usage (fast booting as a result) of resources.

3. Server market is dominated by Linux.

4. The Cloud can be easily managed by Linux.

5. The story is when one cannot kill, one has to embrace it.

Microsoft is lately doing it to survive in the Cloud Market. Its front end is Microsoft but the back end and the work horse in the Cloud is Linux.

Amazing discovery for Microsoft lullabies.

The bottom line is if you have money buy the elegant iPhone. If you are low in your budget or want a second smart phone buy an Android and not a Microsoft (you have to pay for every little thing you download).

Why do you load your smart phone with applications you never use?

The application load slows its speed, especially if the phone is low in RAM..

Ten applications (for me five) is more than enough.

If you are a geek go for twenty (20) or more.

Of course you have Samsung (again Linux derivative) which has an attractive design.

In the meantime, F-Droid free software version of Android is building up its repertoire of software.

Linux is all over the place and Microsoft cannot monopolize, now. I am one of those guys who used Linux exclusively and I pity the guys/girls who pay for a product which is out of date.

The hidden conspiracy was the binding rule of our history.

But it is notable that there is small but significant number of young scientific individuals who did not like this scenario.

Linus Torvalds pioneered this phenomenon.

One must not forget Richard Stallman popularly known as RMS in Linux community.

The Free Software Foundation (FSF) is a non-profit organization founded by Richard Stallman on 4th October 1985 to support the free software movement, which promotes the universal freedom to study, distribute, create, and modify computer software, with the organization's preference for software being distributed under copyleft ("share alike") terms, such as with its own GNU General Public License.

The FSF was incorporated in Massachusetts, USA, where it is also based.

From its founding until the mid-1990s, FSF's funds were mostly used to employ software developers to write free software for the GNU Project. Since the mid-1990s, the FSF's employees and volunteers have mostly worked on legal and structural issues for the free software movement and the free software community.

History of Linux

1991:

The Linux kernel was publicly announced on 25th August by the 21-year-old Finnish student Linus Benedict Torvalds.

1992:

The Linux kernel was relicensed under the GNU GPL.
The first Linux distribution was created.

1993:

Over 100 developers worked on the Linux kernel.
With their assistance the kernel was adapted to the GNU environment, which creates a large spectrum of application types for Linux.

The oldest currently existing Linux distribution, Slackware, was released for the first time.

Later in the same year, the Debian project was established. Today it is the largest community distribution.

It has more than 60,000 applications.

1994:

Torvalds judges all components of the kernel to be fully matured: He released the version 1.0 of Linux.

The XFree86 project contributed a graphical user interface (GUI).

Commercial Linux distribution makers Red Hat and SUSE publish version 1.0 of their Linux distributions.

1995:

Linux was ported to the DEC Alpha and to the Sun SPARC. Over the following years it was ported to an ever greater number of platforms.

This was probably done by Jon Hall.

1996:

Version 2.0 of the Linux kernel was released.

The kernel can now serve several processors at the same time using symmetric multiprocessing (SMP), and thereby became a serious alternative to Unix, which companies would like to have hands on experience.

1998:

Many major companies such as IBM, Compaq and Oracle announced their support for Linux.

The Cathedral and the Bazaar was first published as an essay (later as a book), resulting in Netscape publicly releasing the source code of its Netscape Communicator (web browser suite). Netscape's action and crediting attributed by above essay brought Linux's open source development model to the attention of the popular technical press.

In addition a group of programmers began developing the graphical user interface KDE.

1999:

A group of developers began work on the graphical environment GNOME, destined to become the free replacement for KDE, which at the time, depended on the, then proprietary, Qt toolkit.

During the same year IBM announced an extensive project for the support of Linux.

2000:

Dell announced that it is now the No. 2 provider of Linux based systems worldwide and the first major manufacturer to offer Linux across its full product line.

2002:

The media reported that "Microsoft killed Dell Linux".

2003:

Android, Inc. was founded in Palo Alto, California in October 2003 by Andy Rubin.

2004:

The XFree86 team split up and joined with the existing X standards body to form the X.Org Foundation, which results in a substantially faster development of the X server for Linux.

2005:

The project openSUSE began a free distribution from Novell's community.

Also the project OpenOffice.org introduced the version 2.0 that then started supporting OASIS OpenDocument standards.

2006:

Oracle released its own distribution of Red Hat Enterprise Linux.

Novell and Microsoft announced cooperation for a better interoperability and mutual patent protection.

2007:

Dell started distributing laptops with Ubuntu pre-installed on them.

2009:

Red Hat's market capitalization equaled Sun's.

This was interpreted as the symbolic moment for the "Linux-based economy".

2011:

Version 3.0 of the Linux kernel was released.

2012:

The aggregate Linux server market revenue exceeded that of the Unix market.

2013:

In ten years Google's Linux-based Android claimed 75% of the smart phone market share, in terms of the number of phones shipped.

2014:

Ubuntu claimed 22,000,000 users.

2015:

Version 4.0 of the Linux kernel was released.

Chapter 30

Myths and Conspiracies in Human History

Myths I can excuse but conspiracies I cannot.

It looks like the entire history of mankind is either filled with myth or conspiracies.

No matter, whether it is philosophy, religion, war or power politics, there is a subtle hint that for success or failure, conspiracy had a bigger role to play.

Conspiracies turned events in history for better or for worse.

But when there is a conspiracy in science and medicine, it is outrageous and deplorable.

I have seen that happening in my entire adult career.

Whether it is the story of cholesterol or the saga of coconut oil story or the saturated oil related industry, the underline manifestation was either misinformation or conspiracy.

They promoted polyunsaturated oil or peanut oil as a remedy.

It was a comedy.

Highly reactive bonds with reactive chemicals were introduced.

From heart attacks to degenerative disease to cancer to dementia ensued.

Who benefited.

Drug companies and doctors.

American Heart Association was instrumental in some of the grand conspiracies.

Why I am writing this?

In another 100 years, when our descendants analyze the history, they will look at us, as jokers in history.

This book is for redemption only.

I was caught up in the grand strategy unknowingly and unwittingly.

I believed what was told like "gospel truth" without any critical analysis!

Conspiracies in Science

In mid nineteen sixties in USA there was a lady scientist who postulated that mitochondria probably originated from bacteria and took a symbiotic existence in evolution in eukaryotic cells.

She would have had some good reasons which was deliberately not recorded in the history of science. She was ridiculed by her male colleagues (may be her female colleagues) and she committed (?attempted) suicide. Mind you, I am going by my memory and tried to retrieve documentary proof of this fact but could not.

The book that contained the information is missing!

Mind you mitochondrial DNA was not known then.

They reproduce by binary fission.

Mitochondrial DNA is always inherited by the mother and the few mitochondria of paternal origin (from the sperm) have no significant role.

Quite unknown to the above incident, few years later, I had the same premonition that mitochondria are almost similar to bacteria in size and structure.

My sense was sharp enough never to say this in medical school to my colleagues or to my teachers. But, I dug in and found there are giant mitochondria in glandular adenomas.

I let it rest at that level till I started teaching pathology many years later.

I am very happy, that the concept is readily accepted now but very unhappy a young life was lost or made miserable by being forthcoming in science.

This is one example of conspiracies in science.

There are many of course.

By way of suppressing information or manufacturing bogus facts are well known in some scientific communities.

There is another conspiracy in America where Federal Governments operate. In some federal states they do not allow the teaching of Evolutionary Concept in Biology.

That is a conspiracy by the evangelical politicians!

Charles Darwin's evolutionary theory comes in conflict with God's creation.

So do not teach it to American Students is the Evangelical Message.

In evolutionary studies mitochondrial DNA is very important. They can record changes that occurred million of years apart. There is something called Mitchondrial Genetics that helps in answering evolutionary puzzles.

There is a protein in evolutionary history called heat shock protein. The heat shock protein of bacteria has remarkable resemblance to heat shock protein in us.

Not only mitochondrial DNA but proteins of prokaryotic cells have links with eukaryotic cells.

Genome projects is currently unraveling the evolutionary secrets one by one.

So it is time for us to forget about the God's creation.

Chapter 31

Emerging Science-01

Intent here is "Killing the Big Bang".

Elliptical galaxies

Elliptical galaxies are the most abundant type of galaxies found in the universe.

They make up to 60% of all the galaxies.

They are the oldest.

However, because of their age and dim qualities, they are frequently outshone by younger, brighter collection of stars. Elliptical galaxies lack the swirling arms of their more well known siblings, spiral galaxies.

Instead, they bear the rounded shape of an ellipse.

Spiral galaxies

Spiral galaxies take their name from the winding spiral shape they demonstrate.

30 percent of the galaxies in the universe observed by scientists are spiral galaxies. These twisted collections of stars and gas often have beautiful shapes and are made up of hot young stars.

Most spiral galaxies contain a central bulge surrounded by a flat rotating disk of stars. Made up of older, dimmer stars, the bulge in the center is thought to contain a massive black hole.

The dim light from the older stars can make the bulge difficult to pinpoint, and there are some spirals that lack this characteristic.

Starburst or Irregular Galaxies

They do not fit either category described above.

About 10 percent are irregular galaxies.

They are probably the youngest galaxies as shown by intense activity and new star formation. Astronomers have found evidence of disk of gas between the stars and they are the source of new star formation.

Galaxy Merger

The galaxies merge with each other due to their massive gravitational forces. They are referred to as the car crash victims of the universe. The resulting galaxy merger is violent and dramatic and change the structure and shape of clashing galaxies.

Dark matter within galaxies

There is dark matter amidst the stars in these galaxies and the exact amount of it is not estimated. Now is the time for me to expand on my hypothesis further in reference to globular or elliptical galaxies and starburst galaxies.

In my postulation the dark mater does not sit idle.

Apart from giving the geometrical shape and stability, they are in the process of transformation.

I would like to use the spiral galaxies to explain my dark matter theory.

I propose, that the dark matter by random events turns into matter!

It is the effect of matter, pulling apart "a string of dark matter" from a saturated or solo dark matter hovering at the periphery of a galaxy consisting of matter, containing right amount of the dark matter, probably slightly in excess of the wondering solo dark matter, that trigger the formation of matter real as opposed to invisible dark matter.

This triggers an event similar to the "Big Bang" but in a smaller scale and dimension (mind you this is my alternative theory to counteract the Big Bang).

The dark matter and its force goes into reverse gear and start producing matter in series of step quite invincible to the observer. It just mergers into the background of the galaxy without a big bang. New stars forming in the spirals which are blue in colour is the end result.

Here the gravitational force of the matter (the galaxy) is in fine tune with the dark forces and dark matter.

There are no shooting or exploding stars but it is a maternity home for new stars in formation and the new stars try to defy (the integrated dark force gives the momentum) the gravitational force built into the spirals.

I am trying to explain why there is a spiral.

The dark matter occupying the space (stable form as opposed to the "transforming bit of strings") let this happen gently and it gives the property of expansion in space to operate in tandem with new star formation.

In this scenario Starburst formation is the most likely at an earlier stage of formation of new stars, giving rise to spiral formation at a later stage and still later when the gravitational forces far exceeds the expansion allowed by the dark forces, the elliptical galaxies form.

Unfortunately, I find that the elliptical galaxies are the antithesis to dark matter transforming into matter. I think, deep in the globular galaxies matter changes into another dimension by forming black holes by shear effect of gravity. Accordingly I believe, the globular galaxies may have trace amount of dark matter in its periphery but nothing substantial deep within its center.

Its center is occupied by black hole which is invisible both due to warping of time and space dimensions and the interference of the light emanating from the globular cluster of stars in the outer perimeter.

So my theory of evolution there is a special place for dark matter and dark holes.

In my theory there is no divine intervention but matter and dark matter in transformation, not in chaos but in random fashion.

Matter and dark matter are two sides of the same coin with opposing properties.

One is visible the other is invisible.

Our obsession to matter visible has made us to ignore the dark matter, as substantial as it is, and may be 10 to 100 times more than the matter real.

Our obsession has a blinding effect.

The dark matter is in continuum, with the physical properties of matter and energy, we measure in physics but in a different dimension merging imperceptibly, probably in dimensions above three or four (as seen as "visible and not visible forms", in the newly forming stars in the distant galaxies).

Its wonder is, it needs different type of quantum physics to grasp it. I call it "the humble pie effect" of not knowing the alternative physics.

My bone of contention of Big Bang is, it has no explanation for enormity of dark matter and also dark holes.

Emerging Science-02

This is my alternative theory to counteract the "Big Bang", the Holy One?

I have said, I am uncomfortable with the "Big Bang" and if I do not propose an alternative theory, then my scientific reasoning becomes void. I want to propose a random theory not a unifying one.

It may sound bizarre but it is tangible in theory.

The space is taken into account with its virtual particles.

In this theory of "Whole of the Universe" is taken into consideration, the matter is only the small change.

It is less than 5%.

The dark matter is said to be 25%.

The rest 70% (near enough is the dark energy).

My assumption is that matter which carry energy with charge always move towards a more stable state without charge. Only way for it to make that move is to become dark matter and in that process it releases dark energy and dark particles without charge.

This in fact, makes the space expand.

The matter moving out and universe expanding are two sides of the coin.

It is the virtual reality.

In this theory the matter reduces not by constant shift but by random events. So when the matter is reduced to a critical point, at a particular location, say for example, less than 1%, something intangible has to happen.

There is a mismatch.

Dark energy and dark mater cannot rule the universe.

It has to go into reverse gear, not in a "Big Bang" but in a series of steps. So the dark energy and dark matter is consumed and new matter is formed and the worlds, galaxies and new universes are formed.

There is neither beginning nor end, cyclical phenomena are in existence.

(This theory is akin to Sansara, no beginning no end. I hope no Buddhist take this theory and start preaching Buddhism is scientific and concoct bizarre preachings).

There is only two or three dimensions to this theory but if few more dimensions are mathematically extrapolated, the possibilities are infinite.

The way to test this theory is to test whether matter can be changed to dark matter or dark forces.

This is more feasible since only charged particles are necessary to neutralize the charge not antimatter.

No annihilation is envisaged.

The converse of it is to try change dark matter to matter. Bombarding dark mater to make matter is assumed more difficult due to its stability and its inherent expansion.

In other words contracting (compacting or concentrating space) space is insurmountable.

If that can be achieved "the time travel" actually become a feasible preposition.

This is my alternative theory to counteract the "Big Bang".

There are two big holes in the Big Bang theory.

How come something come from nothing?

It is akin to creation by god.

Unfortunately scientists a century ago had to satisfy the Church before uttering any theories fearing misgiving by the Church. The Church was the authority and it had to ratify. It was a pity.

Second was the inadequate explanation of the dark matter, a constant was proposed but this did no go well with the contemporaries. Why there was so much dark mater was never adequately explained.

The possibility of "Big Crunch" as opposed "Big Bang" was excluded by default.

There is lot of dark matter amidst our galaxy. We know the matter is consumed in the process of emitting protons and slowly but surely matter is exhausting itself.

There is transformation not annihilation.

Is it possible, the dark mater embedded in our galaxy is a byproduct of this transformation?

If the dark matter is expanding we got to seriously consider this probability. If the dark mater is decreasing (in our galaxy) my theory has to be discarded outright.

If dark matter is stable we got to find an appropriate explanation.

How it got trapped amidst the matter of our galaxy needs some explanation.

Black holes of dying stars may be a candidate but black holes by themselves are a mystery to modern physics.

Emerging Science-03

Is it possible to explain the arrangement of the galaxies using zip model of dark matter?

Dark matter cannot be seen and has to be visualized like virtual particles.

What we can measure is either expansion of space (red shift) or contraction of space (blue shift). By visualizing red shift and blue shift and also by superimposition of matter (stars and galaxies) within the dark matter one has to postulate the possible interaction.

For this to manifest we have to postulate matter can be transformed into dark matter and dark matter to matter, vice versa.

That is why I maintain they are two sides of the coin.

Let us visualize the dark matter arrangement.

Like matter attract like matter and they merge (mix) like water.

The force behind is the Dark Force that make virtual dark particles move about freely in space and also expand. Unlike matter that has to compete for its space forming atoms of increasing heavy (from hydrogen to helium to uranium) density dark matter once formed expands seamlessly (since there is no space constraint).

That is why there is abundance of dark matter and its existence is not governed by typical or current physics.

The zipped dark matter may be thin or may be flat.

It can be coiled.

It can be a coiled coil.

I defer using a coiled coil, such as in DNA since mechanics there is physical bonding between unlike particles of different charges reacting with each other, while completing for limited space at subatomic levels.

Dark matter has no space hindrance.

It can be an opened zip at one end with variable length for each limb and the next line of continuity is a stretch (gap) of tight zip and then the dark matter can open into a unscalable unzipped dimension of considerable enormity and the next portion zipped again and forming imaginary loops that may look like coils that lets the matter align in between (Stars and galaxies can align and rotate in one direction. It could give birth to a globular galaxy of high density). The amount of dark matter is less and formation of matter is very limited in globular galaxies.

Some of these globular forms may collapse so much due to their own gravity and then may end up as dark holes.

The unzipped portion would give the high density Starburst or irregular galaxy assuming that the unzipped portion of dark matter would change to matter (hydrogen to begin with) and collapse in space causing the galaxy to become smaller.

The scarcity of the dark matter probably would allow some of the larger galaxies to split off to smaller galaxies with intensely burning stars.

I am at a bit of a fix to explain the spiral galaxies.

Their shape is probably due to the speed of their rotation. They are moving at tremendously high velocity and most likely due to relative lack of dark matter, stars at the periphery try to defy the gravity of the formation and escape to become independent galaxies.

Barred galaxies are probably formed to retard this process.

In that scenario the dark matter gives the stability.

In the same stretch of imagination the barred galaxies are old whereas the spiral galaxies without bars are younger.

How I am going to fit my ZIP theory.

Let me begin with the easy point of interpretation.

In the barred spiral the matter is converted to dark matter to give its stability. Apart from that how it is formed at the beginning of the spiral is difficult to explain. It may be due to the speed of rotation that matter is under huge stress and torn apart to subatomic particles and realigned to form dark matter with dark force.

It is feasible but what is the role of dark matter in the spirals without bars?

There is a halo in the disk and the spiral portion of the galaxies which is where the mysterious dark matter probably is resident. We know it gives the stability and disk shape.

Is the dark matter takes a neutral role not changing into matter and hydrogen stars?

In fact, in my hypothesis, however outrageous it might sound, I am inclined to think, outer spiral is the location where most of the transformation of dark matter to matter is happening. That is the very reason they are trying to (new stars with tremendous new energy) escape from the globular center and later formation of the bar (bar is like the umbilical cord holding on) to counteract the escape.

It is probably the maternity home of new stars with the mother dark matter wants to procreate as matter (stars) visible.

She does want to hide herself in darkness all the time, so to say.

In other words spiral arms are the burning factory of the universe.

How is my zip theory fall into place?

It could be that a ball of zipped dark matter sits in the middle of the disk and extend outwards strings of open unzipped arms and because of the rotation and high speed takes up the appropriate alignment of a spiral.

That is my elaborate explanation.

Emerging Science-04

Zipping and Unzipping Model gives Flexibility and Modularity to Dark Matter

Modularity is the degree to which a system's components may be separated and recombined.

Zipping model can have many dimensions.

It can exist in a string formation or as a wave (vibrations can be in many ways).

When applied to dark matter, the zipped formation can have a globular or twisted arrangement like in the globular galaxies and spiral galaxies.

Because of its open ends of the ZIP it can form network of connections, some very active some docile.

What is more pertinent is that the mathematics of the conformations can be constructed.

(The modularity theorem can be used to establish a connection between elliptic curves and modular forms).

The light warping around a dark matter into arcs and mirror images can be explained by describing dark matter in zipped and or unzipped arrangement at a particular point in space in relation to matter.

Above all it can fit in with my hypothesis of rearrangement of dark matter to form matter and matter to form dark matter in their transformations.

Lot of things fall into place and modified string hypothesis has place in the Zipped Model.

Chapter 32

Astrology, Astrophysics and switching of Time Dimension

Sri-Lankans are well known for time wasting.

Even the nature's call of poor gecko (hoona) is not spared.

With Avurudu (New Year) coming an oblique reference to mundane affairs of conventional time and adopted time is worth pondering.

In physics time is a concept but not something substantial like momentum, gravity and electro-magnetism.

It is concept that is vital for referring the relationship of moving objects.

If there are no moving objects the time would become more abstract than what it is now.

In black holes the time is warped.

Having said that there is no hindrance for anybody in observing the time scale of Sackra when the need arises and the time is right.

He has to add 25 to 30 minutes to Greenwich Mean Time as he pleases.

Because it is a holiday there is no problem for the commercial institutes or the government to allow the people to observe whatever one pleases but when it is made by compulsion by an astrologer it is nothing but arbitrary.

Nobody should force it down the throat by compulsion.

It is antidemocratic.

I should make some observations that made me to take on astrology not as an obsession in my life. Nobody should deny the fact that we are bound to our mother earth and any escape from earth atmosphere is a tremendous stress on our physiology.

On that count if one worships the mother earth it is just one of respect but not of reverence (what is happening in the world today is quite the opposite).

Giving respect to terrestrial bodies other than sun and moon to me seem and exaggeration of this reference of importance to all moving objects in space.

My father was good at mathematics and in his spare time he took interest in astrology and he could make a chart himself without going to an established astrologer.

He had quite a number of English books and he had made some reference to all his siblings.

I was just eight years and quite by accident I happen to glance through my horoscope predictions noted in the book with some phrases underlined in red.

Now I cannot remember any details but one reference I remember even today.

That I would have a fatal accident and would die when 9 years old. This was a terrifying forecast and I feared becoming 9 years old.

Come my nine year birthday I was terrified.

I could not wait till the dawn of the next day after my birthday lest I would be dead. Next day dawned and I remained alive and the ordeal was over.

However, I thought the book made a mistake and it would be the next birthday that I would die.

And it was an impatient year till the tenth birth day and I never told anybody about this fear and silently went through the ordeal.

Come next birthday I was well prepared for it if I live through just another day I would be defeating the Maraya.

I accomplished that and all of a suddenly my fear disappeared into thin air and from that day onwards I feared nobody teachers, principals, bully boys or for that matter anybody who is a body of substance.

Everybody in the village and in the school new my prowess even though I was not heavily built.

Teachers of course encouraged my aptitude and becoming a student leader was automatic and some students feared me more than the teachers. Of course my rivals used underhand tactics like hiring market thugs but even that did not deter me and the thugs also knew I was a different kettle of fish.

I was swift in avoiding dangers and swift in counter ploys.

But come 18 years and successful in the university entrance, I was determined not to take any leadership roles and declined all posts except in sports which I enjoyed (team games only).

However, one lasting effect was I did not take my birth day seriously thereafter.

In my ninth year every day other than birth day was fun lest I did kick the bucket on the next birthday. I enjoyed every bit of the year except birthday.

Now of course everyday is a day of celebration for me.

The second incident was related to my mother.

My mother was symbol of innocence and a lady to the word but she could be easily cheated by even an average vendor. But that was not a problem for me but she could be easily cheated by fortune tellers.

This fact, I was very observant and one day she took me to a far away place where a women would after a trance would utter the fortunes. We went to the "devale" and the due "panduru" was offered and we were waiting in a queue, for some time.

During this waiting period I decided that I am going to teach a lesson to the fortune teller and my mother once and for all.

Finally, my mother went in and I refused to go in.

No sooner she was in.

I bolted.

I had only 10 cents in my hand (that was the bribe for sweets my mother gave me to accompany her in this ordeal for both of us) and that was not enough for the bus fair and I walked nearly seven to eight miles up to our bus station where the 10 cents would take me home.

Once home I hid atop a tree waited till her arrival. She came home in tears and looked for me and and not finding me inside she broke down.

I thought that she had enough for a day and quietly climbed down from the tree and appeared in front of her with a mischievous grin.

She was neither angry nor happy and in her entire life she had never seen a brat like me.

After a time, I opened up and got her to promise that she should never go to a fortune teller and asked her whether the fortune teller told her about my whereabouts.

The answer was in the negative. I am not going to tell the father if you do keep your promise.

She affirmed and this was a secret we kept our entire life and bearing it out now is good for my soul.

Incidentally, this happened in my ninth year.

My father in due course lost interest in astrology on his own accord and was a good reader of Dhamma and Wood House.

In my entire life I have never gone to a "devalaya" and never a "panduru" offered and I had never missed a plane or flight even though I have traveled half the globe.

If the air travel industry adopt the traveling according to astrologers it would be chaotic.

Astrophysics and astrology do not go hand in hand.

However, I have all the respect for the clever mathematicians who invented this myth of astrology to preserve the transfer of mathematical knowledge and astrophysics down many generations to modern day.

If not for that astrophysics would have been lingering in the dark and we may not have discovered new planets.

For me astrology had given me 2 years of agony but if not for that misfortune I would not have discovered my true self potential.

25th of March, 2006

Chapter 33

Laser and Holography and their Uses

Why write about holography in a book of conspiracy?

Simple.

It can be used to create events apparent which are not real life events.

Whats the caveat?

It can be used by the secret service to invent an event such as 9/11 scenario. I won't write anything more about 9/11 but the top brass in the President Bush administration was probably privy to this holographic creation.

What if aliens who fear the man of an impending attack would do?

They with their superior technique would create a "holographic alien presence" using a probe that is manipulated from the mother ship which is out of human vision.

There is no danger to their life.

If we assume Roswell incident did happen and aliens were captured and used as hostages, that is the most likely scenario that would have evolved.

The necessary ground work has to be prepared to make humans start believing in alien presence. They can project extraterrestrial beings (EB) made of holographic form and choreograph, a test event. That is the level they would go for their own safety and they are not ready for the "shoot first and ask question later" American protocol.

That is their antidote. They can easily send a message of "goodwill" to humans, too.

Let us imagine they want to mount an operation to show their superiority in space travel. They can easily stage something like Phoenix Lights.

Equally, assuming if, the Vatican or a peudo-scientist with a religious intent gets hold of this technology and wants to use holography, for the propaganda value.

It is certainly possible to mount an operation simulating "the god has returned from heaven", live.

That is a likely scenario!

I am skeptic and I believe Americans used this technology in "Moon Landing" to hoodwink the world.

In a few years time we will have holographic Studios and "Television" at home.

The technology is developing fast.

What follows below is an introduction to how holography works.

Holography dates from 1947, when British (native of Hungary) scientist Dennis Gabor developed the theory of holography while working to improve the resolution of an electron microscope. Gabor coined the term hologram from the Greek words holos, meaning "whole," and gramma, meaning "message". Further development of this field was hindered during the following decade, as the light sources available at the time were not truly "coherent" (monochromatic or of a single wavelength).

In 1960 Russian scientists N. Bassov and A. Prokhorov and American scientist Charles Towns overcame this barrier. What heralded the progress was the discovery of Laser Light. Its pure, amplified intense light was ideal for making holograms.

A laser is a device that emits light through a process of optical amplification based on the stimulated emission of electromagnetic radiation. The term "laser" originated as an acronym for "Light amplification by Stimulated Emission of Radiation".

In that year the pulsed-ruby laser was developed by Dr. T.H. Maimam. This laser system Leith emitted a very powerful burst of light that lasted only a few nanoseconds (a billionth of a second). In holography a continuous wave of laser is projected to make an image.

Pulsed short burst laser can effectively freeze movement and makes it possible to produce holograms of high-speed events, such as a bullet in flight, and of living subjects.

The first hologram of a person was made in 1967, paving the way for a specialized application of holography: pulsed holographic portraiture.

In 1962 Emmett Leith and Juris Upatnieks of the University of Michigan recognized from their work in side-reading radar that holography could be used as a 3-D visual medium. In 1962 they read Gabor's paper and "simply out of curiosity" decided to duplicate Gabor's technique using the laser and an "off-axis" technique borrowed from their work in the development of side-reading radar. The result was the first laser transmission Upatniekshologram of 3-D objects (a toy train and bird). These transmission holograms produced images with clarity and realistic depth but required laser light to view the holographic image.

Their pioneering work led to standardization of the equipment used to make holograms.

Today, thousands of laboratories and studios possess the necessary equipment;

a continuous wave laser,

optical devices (lens, mirrors and beam splitters) for directing laser light,

a film holder and an isolation table for exposures.

Stability is absolutely essential because movement as small as a quarter wave- length of light during the exposures of a few minutes or even few seconds can completely spoil a hologram. The basic off-axis technique that Leith and Upatnieks developed is still the basis of holographic methodology.

Also in 1962 Dr. Yuri N. Denisyuk from Russia combined holography with 1908 Nobel Laureate Gabriel Lippmann's work in natural color photography. Denisyuk's approach produced a white light reflection hologram which, for the first time, could be viewed in light from an ordinary incandescent light bulb.

Another major advance in display holography occurred in 1968 when Dr. Stephen A. Benton invented white light transmission holography while researching holographic television at Polaroid Research Laboratories. This type of hologram can be viewed in ordinary white light creating a "rainbow" image from the seven colors which make up white light. The depth and brilliance of the image and its rainbow spectrum soon attracted artists who adapted this technique to their work and brought holography further into public awareness.

Benton's invention is particularly significant because it made possible mass production of holograms using an embossing technique. These holograms are "printed" by stamping the interference pattern onto a layered plastic. The resulting hologram can be duplicated millions of times for a few cents apiece. Consequently, embossed holograms are now being used by the publishing, advertising, and banking industries.

In 1972 Lloyd Cross developed the integral hologram by combining white light transmission holography with conventional cinematography to produce moving 3D-three dimensional images. Sequential frames of 2-D motion-picture footage of a rotating subject are recorded on holographic film. When viewed, the composite images are synthesized by the human brain as a 3-D image.

In 70's Victor Komar and his colleagues at the All-Union Cinema and Photographic Research Institute (NIFKI) in Russia, developed a prototype for a projected holographic movie. Images were recorded with a pulsed holographic camera. The developed film was projected onto a holographic screen that focused the dimensional image out to several points in the audience.

Holographic artists have greatly increased their technical knowledge of the discipline and now contribute to the technology as well as the creative process. The art form has become international, with major exhibitions being held throughout the world.

There are two physical phenomena as the principles of the holography: interference and diffraction of light waves.

Holograms are photographs of three dimensional impressions on the surface of moving light waves. In order to make a hologram one need to photograph the individual light waves in a "standing" position or in the interface.

This presents something of a dilemma.

Interference pattern

As we all know, it can be problematic to take a photograph of a quickly moving object. If you've ever had a picture come back blurred from the film laboratory, you know all too well. When a person moves too quickly in a photograph, their image blurs.

Try to imagine the problems associated with trying to photograph a photon.

To start, a light wave moves at the tremendous speed. Thats about 300,000 kilometers per second. Thats more than half way to the moon in a second. Considerably faster than someone's hand waving. In fact, its so fast that the very idea of even capturing it on film would appear impossible.

What we need is a way to stop the photon so it can be photographed and this technique is called interference.

Imagine yourself standing on a small bridge over a of still water. Lets further imagine that you were to drop a pebble into the pond. As it hits the water it creates a circular wave. This wave radiates outwards in an ever growing circular path. We've all seen this. Now, if you drop two pebbles in the water, you would create two circular waves, each of which would grow in size and eventually cross the path of the other wave and then continue on its individual expanding path.

Where the two circular waves cross each other, you might say that they interfere with each other. And the pattern that they make is called an interference pattern.

This is what we call the interference.

Two waves interfering with each other as they cross paths. No permanent impact is left on either wave once it leaves the area of overlap. Each wave looks exactly the same as it did before it crossed the other waves path. Well, maybe its grown a little bit bigger, but that's about it.

So, what's the big deal about interference in that case?

As waves cross paths and interfere, the pattern they make is called a standing wave. It is called a standing wave because it stands still.

And since it stands still, it can be photographed.

This solves the laser device problem of how we can photograph something moving at the speed of light. So, to photograph interference pattern we should use special light source.

Laser light differs drastically from all other light sources, man made or natural, in one basic way which leads to several startling characteristics.

Laser light is coherent.

This means that the light being emitted by the laser is of the same wavelength, and is in phase. Let us make an analogy that could clarify the term coherence.

But it does not answer the bigger question.

Why does light wave stand still?

To understand that, let's envision a photon. If we view it from the side it looks like a sine wave. Now, try to imagine a river whose stream bed lies on a wavy rock formation that looks like a sine wave. This river would be full of rapids.

Although the water in the river is flowing furiously downstream, the pattern of water above the rapids is stationary. You might think of it as a standing wave. The wave energy is flowing through this standing wave without altering it and vice versa. It is just a momentary pattern that the water takes as it passes over a bump.

When two light waves pass through each other each wave acts like a bump to the other. And the result is like rapids of light. The standing wave patterns are stationary even though the light waves energy continues to move.

Transmission hologram

When waves meet they perform addition and subtraction. When two waves of equal size meet at their high points (called crests), they add together to make a wave twice as high at that point. Conversely, where two waves of equal size meet at their low points (call troughs) they add together to become twice as low. And when one wave at its high point meets another wave at its low point they subtract and cancel out.

But it isn't really canceled out in the sense of being destroyed.

Its more a case of there being no light at that spot.

If you follow the wave down its path just a drop further on, it would be meeting the other wave at a different relationship and past that point of interference, once again it will be visible.

Its a situation of infinite possibilities.

Just like the patterns possible as the waves of two pebbles meet in a pond. At any point you may notice that the standing wave pattern has produced a place where the waves have added together to get higher or subtracted to become lower or even just gone flat.

There are few terms that are used to describe the possible encounters.

If the waves add and get higher its called constructive interference. If the waves subtract or cancel altogether its called destructive interference.

Imagine the interference pattern as a fingerprint of the encounter of two individual waves. Each object you make a hologram creates its own interference pattern that identifies it.

Transmission hologram recording

In holography, there are two basic waves that come together to create the interference pattern. First and foremost is the wave that bounces off the object we are making a hologram of. Since it bounces off the object, thereby taking its shape, it is called the object wave. You can't have interference without something to interfere with. So a second wave of light that has not bounced off an object is used to perform this function. It is called the reference wave. When an object wave meets a reference wave creating a standing wave pattern of interference, it is photographed and called a hologram.

Semi-transparent mirror divides laser beam into two beams. The first beam which is called a signal beam, is directed by mirror, expanded by lens and it illuminates object.

The second beam, called a reference beam, is also directed by mirror, expanded by lens and it falls directly onto photoplate. The photoplate registers an interference pattern between the signal beam and the reference beam, reflected by the object.

A transmission hologram appears after an ordinary photo-chemical treatment (hologram of Leith-Upatnieks). If such a hologram is exposed to a laser light beam, you may see a 3-D image of the object.

The transmission hologram does not reconstruct the image in ordinary white light, and it is necessary to copy it to the reflection hologram.

I have only stated the behavior of the light but it can be mixed up with real audio to make a life like holographic performance. The technology has improved tremendously over the years.

Chapter 34

Conspiracies in Politics-01

Why we need a Referendum to get rid of the Need of a Referendum to make Constitutional Effects.

Let me be forthright, I am not in favor of Referendums.

It does not serve any purpose.

There is no scientific basis.

Let me take the New Zealand recent poll to change the National Flag.

Mind you New Zealand had a referendum to make electoral reforms, in early nineteen nineties when I was working there. I was very busy and I had very little interest on electoral systems including proportional representation. People of New Zealand are very nice and down to earth. Whether it is cricket or politics or science they have a passion in what ever they do, unlike citizens in United Kingdom.

They take referendums seriously.

United Kingdom people are generally grumblers.

After so much time and money wasted there was a big NO for change of the flag.

That example alone is enough to repudiate its validity.

The British opted out of European Union by a referendum. It looks like the politicians cannot make intelligent decision once in power and leave it to the voter and waste colossal amount of money and time too.

We had our own shameful referendum to extend the parliament by Old Fox's methodology.

Then we had a lady Prime Minister extending the parliament by two years with two third majority.

Both were ugly incidents in our political history.

Just imagine former president comes into power with strange bed fellows like Vasu (Nanayakkara). Then they will get together and have a referendum to make it possible for the term of office to be increased to 15 years.

Additionally they will make, no bar for number of "times" in office (twice currently). The motive is to enjoy the political power for life and no chance for alternative views.

What they need is 51% to 49% division in the final voting count.

To me that scenario is uglier than the stated referendums above.

We are going through the process of formalizing a new constitution. Suppose at a referendum the constitution is rejected (It is more than likely present rulers will lose).

All the good work that preceded will be down the drain.

The crafty last ex-president knows this very well and that is why he is asking the presidency should be abolished in double quick time.

He has two or three strategies.

1. One is to go for the abolition (at referendum) of the presidency and claim a birth for Prime Minister post in five years time.

2. The second strategy is to derail the new constitution by going for the kill at the referendum.

The gray area of not having a president and no effective Prime Minister with power.

That is the worst scenario.

3. The third possibility is to do a President Clinton Method.

Put his lady as the next presidential candidate and win and make a mockery of two term restriction in office.

He will be running the day to day affairs.

All these are possible scenarios in this blessed country.

That is why I say we should abolish the need for a referendum before we set about making a new constitution.

I am also against (except for some specified reasons) the need for, two thirds majority for making, substantial changes to the present constitution.

It is the lessor of the evils in constitutional making.

I have no interest in politics.

My interests are scientific in nature and my questioning here are based on scientific thinking and not political or a bizarre proclamation made by "a medium" or " a channel".

I hope the young and the old should take cognizant of the probable political scenarios in constitutional making.

Conspiracies in Politics-02

Franchise a mathematical marvel or an aberration of Sri-Lankan Polity

In mathematics only three words tend to interest me.

They are the zero, the infinity and the range of values that abstracts everything else in between.

In statistics which I am always skeptical about, where figures can be manipulated to show the desired result instead of the underlying inherent tendency, the observer can make mistakes in predicting the human behavior, at a given time.

Few words I tend to regard influential in my thinking are the distribution, the tendency and skewing either to the right or to the left out of the many useless terms of insignificance.

If I may explain what they mean in simple terms, the normal distribution is where the range of values (ideas) are distributed graphically like a bell (with a smooth curve on it surface).

In this distribution 68% falls within the centre (the majority fall within this) and 95% falls within the overall bell (shape) distribution and only 5% falls outside the "graphic bell". If anything falls within this 5% area (which is outside the 95% normal distribution) it is taken as an exception to the rule and something remarkable or abnormal feature is declared.

In different distribution the bell shape may vary from a tall bell to a fat broad bell. When range of values (ideas) is narrow the bell is narrow and when there is wide range of values the bell becomes broader and looks fat.

Any opinion or idea or value taking this shape is called a normal tendency and distribution.

But in political terms the opinion / opinions is / are shifted to one side (either to the right or to the left) and this shifting either to the right or left is called skewing.

This skewing is not accepted as normal in statistical terms but in real life this skewing has known for its presence almost instantaneously especially in political dealing and opinion making.

The use of statistics is not universal in politics but a politicians might use these data to either procrastinate on a particular issue or take political advantage with selfish gains. I think our former president JRJ was led to believe the electoral statistics and made sweeping changes to the electoral system.

Whether it is, Chandrika or Ranil or Mahinda Chintanaya or for that matter NGO Chinthanaya or Jathika Chinthanaya is concerned they never fall into normally accepted distribution of the majority of the people (voter and souls not interested in voting).

The minority rules the opinion and the majority has no opinion left to express.

This is true in economics too.

The rich makes the decisions and the poor takes the leftovers and dhanas on the big Poya days.

What is significant in statistical sense is, that the international community exaggerates this skewing effect in the third world countries since their eyes are congenitally skewed and can never be treated in medical sense.

When it comes to democratic exercises the Sri-Lankan polity is permanently skewed beyond any redemption after the JRJ's mathematical exercise of constitutional changes of insane dimensions.

Why I say so, is open to discussion by any sane Sri-Lankan (most the sane ones have left the shores and only the insane ones-some with split minds - or already retired from active service, who return to this island after long stay in the West to avoid the extremes of weather there) if there is any left in this country.

First insane observation was that having perused the election results from independence to 1977 (he did not realize that if not for the first past the post system he would not have got - more than the two third of the majority) that the UNP had a rock solid block vote. First insane inference following that was that if the electoral system is tampered with the UNP and the ruling class can be in power eternally and maintain the skewed distribution of political power.

He never realized if the power slips to the other side they would maintain this skewed distribution for donkey's years by hook or by crook. This is what happened under SLFP and they remained in power for 21 years to 17 years of UNP.

Second insane observation was that all the candidates over a short period of time have lost their margin of majority.

The remedy (inference) was fantastically insane.

Each time a voter casts a vote multiply it by a factor of three so that the candidate's popularity is exaggerated by 300%. This was a fantastic ploy to hoodwink the average man on the street who has not got an average understanding in arithmetic except for counting the fingers. For example a man who gets about 2250 votes which in actual fact is only 750 voters who would have exercised the franchise or their will (barring rigging which can be substantial in a country where a dead man can vote - my father in law voted from heaven in one of the presidential elections). The irony is that these peoples representatives get their images bloated out of proportion (none of them had any image before becoming representatives and our media bloat them further by reporting their act of omissions) to the needs of the poor masses (physical bloating an independent issue - none of them look healthy specimen to emulate the feat of our Commonwealth Games Gold Medalist from Polonnaruwa). The third observation was that when the Kings were ruling this country, the people were better off and replacing the Kings / or Queens by Presidential candidate with only 50% of the vote is the miracle cure for our failures in governance,in the past. Little he realized that only essential road to success is broken promises and damn lies-pathological in dimension. An aberration par excellence is that when 50% is not cast to engineer a second count. Unfortunately 49% of the voters will not have a King / or Queen of their choice.

Conspiracies in Politics-03

Verse 324, 320 and 321 of Dhammapada

Parijinna Brahmanaputta Vatthu
Dhanapalo nama kunjaro
katukabhedano dunnivarayo
baddho kabalam na bhunjati
sumarati nagavanassa kunjaro.

The elephant called Dhanapala, in severe must and uncontrollable, being in captivity, eats not a morsel, yearning for his native forest (i.e., longing to look after his parents).

Aham nagova sangame
capato patitam saram
ativakyam titikkhissam
dussilo hi bahujjano.

As an elephant in battlefield withstands the arrow shot from a bow, so shall I endure abuse. Indeed, many people are without morality.

<u>*Dantam nayanti samitim*</u>
<u>*dantam raja' bhiruhati*</u>
<u>*danto settho manussesu*</u>
<u>*yo' tivakyam titikkhati.*</u>

Only the trained (horses and elephants) are led to gatherings of people; the King mounts only the trained (horses and elephants). Noblest among men are the tamed, who endure abuse.

The Elephant Corridor

It is a myth.

Yala, Wagamuwa, Wilpattu and Uda Walawe are not connected by a corridor for the elephants to move freely, since the rest of the land is densely populated and there is not even a hypothetical corridor.

My calculation is that not even one percent of the land is reserved for the elephants.

If, I assume 10% (ten) of the land is allocated in a hypothetical space on this island as a zoological garden and all the elephants are relocated inside it for the people to see them as exhibits, and only 0.1% of the total population come to the zoo, to see them in one go (again hypothetical), there will be one man or woman standing every 10 feet (3.3 meters) of the perimeter and no elephant can move out without touching a human being.

If we get 1% to visit the zoo, one can stand and won't be able to stretch the hands without pushing and shoving.

Elephant will have to trample 10 of them to get out of the perimeter and see the man's world, which is fully developed.

One may wonder, why I used 10% which is high land mass in commercial concepts.

It is very simple.

My estimate is that there should be 35% (it is less than 25% and we will be like Dubai in about 50 years of development from now and then we might even have to import water) of total land that should be reserved for rain forest, if we are to maintain at least 10% of the perennial rivers.

So if we roughly have 30% for the rain forest, elephants should take the major share of at least one third of it which will be 10%.

That is why I state that there is not even 1% of land left for our elephants.

So the natural selection will let 500 elephants to survive in this land and in that estimate it is really an endangered species. Stating that we have 5000 elephants in the forest is a mythical figure concocted by our Mega Media Men (M.M.M) to satisfy the politicians. I think each elephant is counted ten (10) times in making that mythical figure, sometimes by reputed NGO's.

Development of the man's world is also a big myth in the same stretch of imagination, since only a very few with patronage will be able to enjoy it.

I have made the calculation below confusing and difficult so that politicians won't be able to understand my logic of it but the figures are below for one to work out.

Human Population density in Sril-Lanka is averaged out but in reality it is much more denser than that because people congregate on socio-economic reasons.

The population density in Gamphaha is more than Duka and Calcatta.

Kandy has 10 times the population density of Gaza Strip.

Beauty is nobody talks about population control.

I have used the total population projected in another 10 to 50 years for my calculation since total square miles is roughly 25,000.

1000 in a square mile

400 in a Square Kilometer

400 x 2.4=1000

Reserve only 10% 2500 Sq Miles(6500 Sq. Meters)

0.1 of population 250,000

Every 10 feet or 3.3 meters there is a man/woman standing at the perimeter of land of 10% area.

Location   Indian Ocean   Coordinates   7°N 81°E

Area   65,610 km2 (25,332 sq miles)

Area rank   25th

Coastline   1,340 km (833 miles)

Highest elevation   2,524.13 m (8,281.27 ft)

Highest point   Pidurutalagala

Water bodies   870 km² of water =400 square miles (only about 1%)

(If Water body is 1.5 % it equals to 1000 Km)

Population density   1000 / sq miles

Population density   400 / km³

Hypothetical Elephants   population is 1000 and it is exaggerated to 5000 by local media giants (Biggest lie in this country).

Reserve   2500 Sq miles 50x50=2500;10 feet

Reserve   6500 Km 80x80=6400; 3.3 meters.

Epilogue

Henry Steel Colonel Olcott

Colonel Henry Steel Olcott (2nd August 1832 – 17th February 1907) was an American military officer, journalist, lawyer and the co-founder and first President of the Theosophical Society.

Olcott was the first well known American of European ancestry to make a formal conversion to Buddhism. His subsequent actions as president of the Theosophical Society helped create a renaissance in the study of Buddhism. Olcott is considered a Buddhist modernist for his efforts in interpreting Buddhism through a Westernized lens.

Olcott was a major revivalist of Buddhism in Sri Lanka and he is still honored in Sri Lanka for these efforts. Olcott has been called by Sri Lankans "one of the heroes in the struggle of our independence and a pioneer of the present religious, national and cultural revival".

Theosophical Society

From 1874 on, Olcott's spiritual growth and development with Blavatsky and other spiritual leaders would lead to the founding of the Theosophical Society. In 1875, Olcott, Blavatsky, and others, notably William Quan Judge, formed the Theosophical Society in New York City, USA. Olcott financially supported the earliest years of the Theosophical Society and was acting President while Blavatsky served as the Society's Secretary.

In December 1878 they left New York in order to move the headquarters of the Society to India. They landed at Bombay on February 16th, 1879. Olcott set out to experience the native country of his spiritual leader, the Buddha. The headquarters of the Society were established at Adyar, Chennai as the Theosophical Society Adyar, starting also the Adyar Library and Research Centre within the headquarters.

I forget the exact motives of Olcott and I believe his outlook was much broader but with resurgence of Buddhist revival in Ceylon, his original ideas metamorphosed to a single theme.

His broader inquiry stretched into to religion, philosophy and science, not spirituality alone. But his followers in Ceylon put fullstops to science and philosophy.

Helena Blavatsky

Helena Petrovna Blavatsky (Russian; 12th August 1831 – 8th May 1891) was a Russian occultist, spirit medium, and author who co-founded the Theosophical Society in 1875. She gained an international following as the leading theoretician of Theosophy, the esoteric movement that the society promoted.

Born into an aristocratic Russian-German family in Yekaterinoslav, Ukraine, Blavatsky traveled widely around the Russian Empire as a child. Largely self-educated, she developed an interest in Western esotericism during her teenage years.

According to her later claims, in 1849 she embarked on a series of world travels, visiting Europe, the Americas, and India.

She alleged that during this period she encountered a group of spiritual adepts, the "Masters of the Ancient Wisdom", who sent her to Shigatse, Tibet, where they trained her to develop her own psychic powers.

Both contemporary critics and later biographers have argued that some or all of these foreign visits were fictitious, and that she spent this period in Europe.

By the early 1870s, Blavatsky was involved in the Spiritualist movement, although defending the genuine existence of Spiritualist phenomena, she argued against the mainstream Spiritualist idea that the entities contacted were the spirits of the dead.

Relocating to the United States in 1873, she befriended Henry Steel Olcott and rose to public attention as a spirit medium, attention that included public accusations of fraudulence.

In New York City, Blavatsky co-founded the Theosophical Society with Olcott and William Quan Judge in 1875. In 1877 she published Isis Unveiled, a book outlining her Theosophical world view.

Blavatsky described Theosophy as "the synthesis of science, religion and philosophy", proclaiming that it was reviving an "Ancient Wisdom".

In 1880 she and Olcott moved to India, where the Society was allied to the Arya Samaj, a Hindu reform movement. That same year, while in Ceylon she and Olcott became the first Westerners to officially convert to Buddhism.

Although opposed by the British administration, Theosophy spread rapidly in India but experienced internal problems after Blavatsky was accused of producing fraudulent paranormal phenomena.

Amid ailing health, in 1885 she returned to Europe, there establishing the Blavatsky Lodge in London. Here she published The Secret Doctrine, a commentary on what she claimed were ancient Tibetan manuscripts, as well as two further books, The Key to Theosophy and The Voice of the Silence.

She died of influenza in the home of her disciple and successor, Annie Besant.

Blavatsky was a controversial figure during her lifetime, championed by supporters as an enlightened guru and derided as a fraudulent charlatan and plagiarist by critics. Her Theosophical doctrines influenced the spread of Hindu and Buddhist ideas in the West as well as the development of Western esoteric currents.

Annie Besant

Annie Besant (1st October 1847 – 20th September 1933) was a prominent British socialist, theosophist, women's rights activist, writer and orator and supporter of Irish and Indian self-rule.

In 1867, Annie at age 20, married Frank Besant, a clergyman, and they had two children, but Annie's increasingly anti-religious views led to a legal separation in 1873. She then became a prominent speaker for the National Secular Society (NSS) and writer and a close friend of Charles Bradlaugh.

In 1877 they were prosecuted for publishing a book by birth control campaigner Charles Knowlton. The scandal made them famous, and Bradlaugh was elected M.P. for Northampton in 1880.

She became involved with union actions including the Bloody Sunday demonstration and the London match-girls strike of 1888. She was a leading speaker for the Fabian Society and the Marxist Social Democratic Federation (SDF). She was elected to the London School Board for Tower Hamlets, topping the poll even though few women were qualified to vote at that time.

In 1890 Besant met Helena Blavatsky and over the next few years her interest in theosophy grew while her interest in secular matters waned. She became a member of the Theosophical Society and a prominent lecturer on the subject. As a part of her theosophy related work, she often traveled to India.

In 1898 she helped establish the Central Hindu College and in 1922 she helped establish the Hyderabad (Sind) National Collegiate Board in Mumbai, India.

In 1902, she established the first overseas Lodge of the International Order of Co-Freemasonry, Le Droit Humain. Over the next few years she established lodges in many parts of the British Empire.

In 1907 she became president of the Theosophical Society, whose international headquarters were in Adyar, Madras, (Chennai).

She also became involved in politics in India, joining the Indian National Congress. When World War I broke out in 1914, she helped launch the Home Rule League to campaign for democracy in India and dominion status within the Empire. This led to her election as president of the India National Congress in late 1917.

In the late 1920s, Besant traveled to the United States with her protégé and adopted son Jiddu Krishnamurti, whom she claimed was the new Messiah and incarnation of Buddha. Krishnamurti rejected these claims.

After the war, she continued to campaign for Indian independence and for the causes of theosophy, until her death in 1933.

Buddhism in America

Not all Americans are bad. There are lot of good Americans. What I am against are the American Scientists who conspire for the last 70 years.

It is time I should say good about Americans.

I have met few of them and there are many like Dr. Steven M. Greer. Unlike British, Americans are fabulously rich and spend time and money to explore the unknown, including UFOs.

Whereas the British will hold onto their prejudices in their entire life. Even though, yesteryear British scholars had revived and reviewed Buddhist scriptures (Francis Story and Rhys Davies) after the visit of Colonel Henry Steel Olcott to Ceylon, there is none left in Ceylon and UK.

Rhys Davies

Thomas William Rhys Davids, FBA (12th May 1843 – 27th December 1922) was a British scholar of the Pāli language and founder of the Pali Text Society.

Thomas William Rhys Davids was born in England, at Colchester in Essex, the eldest son of a Congregational clergyman from Wales, who was affectionately referred to as the Bishop of Essex. His mother, who died at the age of 37 following childbirth, had run the Sunday school at his father's church.

He took an active part in founding the British Academy and London School for Oriental Studies. He studied for the bar and briefly practiced law, though he continued to publish articles about Sri-Lankan inscriptions and translations, notably in Max Müller's (he was a French Scholar) monumental Sacred Books of the East.

From 1882 to 1904 Rhys Davids was Professor of Pali at the University of London, a post which carried no fixed salary other than lecture fees. In 1905 he took up the Chair of Comparative Religion at the University of Manchester.

Rhys Davids attempted to promote Theravada Buddhism and Pāli scholarship in Britain. He actively lobbied the government (in co-operation with the Asiatic Society of Great Britain) to expand funding for the study of Indian languages and literature, using numerous arguments over how this might strengthen the British hold on India. He gave "Historical Lectures" and wrote papers advancing a racial theory of a common "Aryan" ethnicity amongst the peoples of Britain,

Sri Lanka, and the Buddha's own clan in ancient times. These were comparable to the racial theories of Max Müller, but were used to a different purpose.

Rhys Davids claimed that Britons had a natural, "racial" affinity with Buddhist doctrine.

This part of Rhys Davids' career is controversial.

His scholarly enterprise was terminated by a bogus investigation by the British.

British Conspiracy

He never renunciate his faith but promoted study of Buddhism. That was his downfall. Rhys Davids' civil service career and his residence in Sri Lanka came to an abrupt end. Personal differences with his superior, C. W. Twynham, caused a formal investigation, resulting a tribunal and Rhys Davids' dismissal for misconduct. A number of minor offenses had been discovered, as well as grievances concerning fines improperly exacted both from Rhys Davids' subjects and his employees.

With his death study of Buddhism ended in Great Britain.

Unlike, the British, the American continue to study Buddhism up to this date. A considerable number of Americans practice Buddhist Way of Life without enunciating their own faith (for political reasons).

Buddhism is thriving in America!

They have contributed a lot silently.

Looking for absolute truth

I never joined a club or secret society in my entire life. Club mentality was an alien theme to me, for the simple reason of confining myself to a narrow segment of free inquiry. In practice it closes all the avenues of investigation of new and emerging knowledge, especially science.

Having returned from united Kingdom with lot of new ideas, I was invited to a symposium organized by a spiritual organization. I grabbed the opportunity and said it should include all three disciplines, philosophy, religion and science.

No politics by conviction.

I took the responsibility of organizing it but did not have a clue how to approach the theme philosophy.

Luckily, I knew Monte Gopallawa, mostly indirectly (my sisters batch) who had read philosophy in his undergraduate studies. He was the Governor of the Central Province and was easy to approach. Through his offices I managed to organize a seminar and an introduction to Professor A.D.P Kalansuriya.

We decided to have a "short weekly discussion" in the University of Peradeniya and invited few guys and girls. In the first discussion we had about 10 and in one of the subsequent discussions, one guy (I did not know him) in a national dress, who came from the Buddhist background, started arguing on a flimsy point slanted towards Buddhism, off the track of professors theme by many a mile.

Professor in his inimitable way, tried to put him on the track with very short doses of philosophy. This idiot went on in circles and after waiting awhile, annoyed to the brim, I blasted this guy and told him, we came to listen to philosophy not Buddhism (this idiot's interpretation, not Buddha's) and you better get out of this room or I will throw you out physically.

I got up from my seat and showed him, I meant what I said. He with his followers left the room and there was one other guy (may be my very young brother in law) beside me left in the room.

Professor had a bright smile in his face and after a pause said these guys cannot understand a simple term in philosophy. That is the problem I am faced with.

We continued to have the discussions as planned and in subsequent discussions I was the only guy left.

Prof ADP Kalansuriya

The above incident brought us together and made our encounters frequent. He was probably the last true philosopher we had. He had few medical problems and never sought any advice from me and within a short time passed away, cutting short our acquaintance.

His thesis in UK was accepted without a hitch, in early sixties, not in any way slanted to Buddhism. I was lucky to have receive a free copy of his last book, related to language games in philosophy. It was very heavy stuff in content and English terms, which I had to refer the dictionary to grasp his thoughts.

That was the only book in my life I read one chapter at a time (several times) and never proceeded to next chapter without getting hold of his dialectical and eclectically themes.

Listening to him over six to ten close door sessions made me well prepared.

One of his genuine attempts was to get us out of the rigid and prejudiced "inside the box" views. He said and reiterated, even, philosophers are trapped in this very same dilemma. He encouraged us to think "out of the box" and showed us the way out of a dogma.

I am very thankful to him but he is no more.

He succumbed to a medical misadventure just like Monte Gopallawa (Monte died before him) his only and the last student.

Coming back to "Looking for absolute truth", I organized the symposium and invited him for a short introduction to philosophy in Sinhala.

I of course dished out the science theme and after a few introductory remarks made the talk into a discussion forum, "a question and short answers", knowing very well that the audience was not ready for advanced science topics.

That was the last time, I addressed a public lecture 15 years ago. I try my level best to avoid any, simply because the general public want to be within the "inside the box view".

Getting them out is difficult philosophically.

Authors Note

I have to make a confession.

I have not "fact checked" most of the claims attributed to the authors mentioned in this book.

In conspiracy or speculation, "fact checking" is humanly impossible. Errors are built into the system by repetition and the difference between "fact" and "fiction" becomes almost imperceptible.

In this context, some of the films made on verifiable facts, inadvertently become fiction due to inflation or exaggeration.

It works both ways, a fact becomes a fiction and a fiction becomes a fact.

In this scenario conspiracy thrives.

I am one who is averse to lies or deception.

I want scientists to be "bona fide" true gentlemen / ladies and when I find the "so called scientists", deliberately lie for personal or corporate gains, it is obnoxious.

Unfortunately, I have seen doctors working in alliance with representative from pharmaceutical companies, use lies, deception, partial disclosure of side effects and exaggerating flimsy benefits.

Having made unwitting entry into fields foreign to my own, as a preparation for this book, now I am convinced conspiracy in theory is universal, an Achilles Heel to scientific verification.

Wikipedia had been a major source and I am thankful to all who contribute to Wikipedia.

But any errors of commission and omission are all mine.

I have placed many personalities in this book not knowing their direct or indirect contribution to either science or conspiracy.

I had no personal choices, in the selections I have made, in this exercise and I may have inadvertently, left out many important personalities.

I present this book to those personalities left out unwittingly, but nevertheless, suffered ridicule and reprimand for their genuine interest and dedication in seeking the truth.

I hope this book is an inspiration for those who seek transparency in science, governance and politics.

Asokaplus

www.ingramcontent.com/pod-product-compliance
Lightning Source LLC
Chambersburg PA
CBHW080649190526
45169CB00006B/2047